# Mathematical Surprises

Mordechai Ben-Ari

# Mathematical Surprises

 Springer

Mordechai Ben-Ari
Department of Science Teaching
Weizmann Institute of Science
Rehovot, Israel

ISBN 978-3-031-13565-1          ISBN 978-3-031-13566-8   (eBook)
https://doi.org/10.1007/978-3-031-13566-8

Mathematics Subject Classification (2020): 00-01, 51Nxx, 05-01, 00A07

This Springer imprint is published by the registered company Springer Nature Switzerland AG
The registered company address is: Gewerbestrasse 11, 6330 Cham, Switzerland

# Foreword

> If everyone were exposed to mathematics in its natural
> state, with all the challenging fun and surprises that that
> entails, I think we would see a dramatic change both in the
> attitude of students toward mathematics, and in our
> conception of what it means to be "good at math."
> Paul Lockhart

> I'm really hungry for surprises because each one makes us
> ever-so-slightly but substantially smarter.
> Tadashi Tokieda

Mathematics, when appropriately approached, can provide us with plentiful pleasant surprises. This is confirmed by a Google search of "mathematical surprises," which, surprisingly, yields almost half a billion items. What is a surprise? The origins of the word trace back to Old French with roots in Latin: "sur" (over) and "prendre" (to take, to grasp, to seize). Literally, to surprise is to overtake. As a noun, surprise is both an unanticipated or bewildering event or circumstance, as well as the emotion caused by it.

Consider, for example, an extract from a lecture by Maxim Bruckheimer[1] on the Feuerbach circle: "Two points lie on one and only one straight line, this is no surprise. However, three points are not necessarily on one straight line and if, during a geometrical exploration, three points 'fall into' a straight line, this is a surprise and frequently we need to refer to this fact as a theorem to be proven. Any three points not on a straight line lie on one circle. However, if four points lie on the same circle, this is a surprise that should be formulated as a theorem. ... Insofar as the number of points on a straight line is larger than 3, so is the theorem the more surprising.

---

[1] Maxim Bruckheimer was a mathematician who was one of the founders of the Open University UK and Dean of its Faculty of Mathematics. He was Head of the Department of Science Teaching at the Weizmann Institute of Science.

Likewise, insofar as the number of points lying on one circle is larger than 4, so is the theorem the more surprising. Thus, the statement that for any triangle there are nine related points on the same circle ... is very surprising. Moreover, in spite of the magnitude of the surprise, its proof is elegant and easy."

In this book Mordechai Ben-Ari offers a rich collection of mathematical surprises, most of them less well known than the Feuerbach Circle and with sound reasons for including them. First, in spite of being absent from textbooks, the mathematical gems of this book are accessible with just a high school background (and patience, and paper and pencil, since fun does not come for free). Second, when a mathematical result challenges what we take for granted, we are indeed surprised (Chaps. 1, 13). Similarly, we are surprised by: the cleverness of an argument (Chaps. 2, 3), the justification of the possibility of a geometric construction by algebraic means (Chap. 16), a proof relying on an apparently unrelated topic (Chaps. 4, 5), a strange proof by induction (Chap. 6), new ways of looking at a well-known result (Chap. 7), a seemingly minor theorem becoming the foundation of a whole field of mathematics (Chap. 8), unexpected sources of inspiration (Chap. 9), rich formalizations emerging from purely recreational activities such as origami (Chaps. 10–12). These are all different reasons for the inclusion of the pleasant, beautiful and memorable mathematical surprises in this lovely book.

So far I have addressed how the book relates to the first part of the definition of surprise, the cognitive rational reasons for the unexpected. As to the second aspect, the emotional aspect, this book is a vivid instantiation of what many mathematicians claim regarding the primary reason for doing mathematics: it is fascinating! Moreover, they claim that mathematics stimulates both our intellectual curiosity and our esthetic sensibilities, and that solving a problem or understanding a concept provides a spiritual reward, which entices us to keep working on more problems and concepts.

It has been said that the function of a foreword tell readers why they should read the book. I have tried to accomplish this, but I believe that the fuller answer will come from you, the reader, after reading it and experiencing what the etymology of the word surprise suggests: to be overtaken by it!

*Abraham Arcavi*

# Preface

Godfried Toussaint's article on the "collapsing compass" [50] made a profound impression on me. It would never have occurred to me that the modern compass with a friction joint is not the one used in Euclid's day. In this book I present a selection of mathematical results that are not only interesting, but that surprised me when I first encountered them.

The mathematics required to read the book is secondary-school mathematics, but that does not mean that the material is simple. Some of the proofs are quite long and require that the reader be willing to persevere in studying the material. The reward is understanding of some of the most beautiful results in mathematics. The book is not a textbook, because the wide range of topics covered doesn't fit neatly into a syllabus. It is appropriate for enrichment activities for secondary-school students, for college-level seminars and for mathematics teachers.

The chapters can be read independently. (An exception is that Chap. 10 on the axioms of origami is a prerequisite for Chaps. 11, 12, the other chapters on origami.) Notes relevant to all chapters are given below in list labeled Style.

## What Is a Surprise?

There were three criteria for including a topic in the book:

- The theorem surprised me. Particularly surprising were the theorems on constructibility with a straightedge and compass. The extremely rich mathematics of origami was almost shocking: when a mathematics teacher proposed a project on origami, I initially turned her down because I doubted that there could be any serious mathematics associated with the art form. Other topics were included because, although I knew the results, their proofs were surprising in their elegance and accessibility, in particular, Gauss's purely algebraic proof that a regular heptadecagon can be constructed.

- The material does not appear in secondary-school and college textbooks, and I found these theorem and proofs only in advanced textbooks and in the research literature. There are Wikipedia articles on most of the topics, but you have to know where to look and the articles are often outlines.
- The theorems and proofs are accessible with a good knowledge of secondary-school mathematics.

Each chapter concludes with a paragraph *What Is the Surprise?* which explains my choice of the topic.

## An Overview of the Contents

Chapter 1 presents Euclid's proof that any construction that is possible with a fixed compass is possible with a collapsing compass. Many proofs have been given, but, as Toussaint shows, most are incorrect because they depend on diagrams which do not always correctly depict the geometry. To emphasize that one must not trust diagrams, I present the famous alleged proof that every triangle is isoceles.

Over the centuries mathematicians unsuccessfully sought to trisect an arbitrary angle (divide it into three equal parts) using only a straightedge and compass. Underwood Dudley made a comprehensive study of trisectors who find incorrect constructions; most constructions are approximations that are claimed to be accurate. Chapter 2 starts by presenting two of these constructions and developing the trigonometric formulas showing that they are only approximations. To show that trisection using just a straightedge and compass is of no practical importance, trisections using more complex tools are presented: Archimedes's *neusis* and Hippias's *quadratrix*. The chapter ends with a proof that it is impossible to trisect an arbitrary angle with a straightedge and compass.

Squaring a circle (given a circle construct a square with the same area) cannot be performed using a straightedge and compass, because the value of $\pi$ cannot be constructed. Chapter 3 presents three elegant constructions of close approximations to $\pi$, one by Kochański and two by Ramanujan. The chapter concludes by showing that a quadratrix can be used to square a circle.

The four-color theorem states that it is possible to color any planar map with four colors, such that no countries with a common boundary are colored with the same color. The proof of this theorem is extremely complicated, but the proof of the five-color theorem is elementary and elegant, as shown in Chapter 4. The chapter also presents Percy Heawood's demonstration that Alfred Kempe's "proof" of the four-color theorem is incorrect.

How many guards must be employed by an art museum so that all the walls are under constant observation by at least one guard? The proof in Chapter 5 is quite clever, using graph coloring to solve what at first sight appears to be a purely geometrical problem.

Chapter 6 presents some lesser-known results and their proofs by induction: theorems on Fibonacci numbers and Fermat numbers, McCarthy's 91 function, and the Josephus problem.

Chapter 7 discusses Po-Shen Loh's method of solving quadratic equations. The method is a critical element of Gauss's algebraic proof that a heptadecagon can be constructed (Chapter 16). The chapter includes al-Khwarizmi's geometric construction for finding roots of quadratic equations and a geometric construction used by Cardano in the development of the formula for finding roots of cubic equations.

Ramsey theory is a topic in combinatorics that is an active area of research. It looks for patterns among subsets of large sets. Chapter 8 presents simple examples of Schur triples, Pythagorean triples, Ramsey numbers and van der Waerden's problem. The proof of the theorem on Pythagorean triples was accomplished recently with the aid of a computer program based on mathematical logic. The chapter concludes with a digression on the ancient Babylonians' knowledge of Pythagorean triples.

C. Dudley Langford observed his son playing with colored blocks and noticed that he had laid them out in an interesting sequence. Chapter 9 presents his theorem on the conditions for such a sequence to be possible.

Chapter 10 contains the seven axioms of origami, together with the detailed calculations of the analytic geometry of the axioms, and characterizations of the folds as geometric loci.

Chapter 11 presents Eduard Lill's method and the origami fold proposed by Margharita P. Beloch. I introduce Lill's method as a magic trick so I won't spoil it by giving details here.

Chapter 12 shows that origami can perfom constructions not possible with a straightedge and compass: trisecting an angle, squaring a circle and constructing a nonagon (a regular polygon with nine sides).

Chapter 13 presents the theorem by Georg Mohr and Lorenzo Mascheroni that any construction with a straightedge and compass can be performed using only a compass.

The corresponding claim that a straightedge only is sufficient is incorrect, because a straightedge cannot compute lengths that are square roots. Jean-Victor Poncelet conjectured and Jakob Steiner proved that a straightedge is sufficient, provided that there exists a single fixed circle somewhere in the plane (Chap. 14).

If two triangles have the same perimeter and the same area must they be congruent? That seems reasonable but it turns out not to be true, although it takes quite a bit of algebra and geometry to find a non-congruent pair as shown in Chap. 15.

Chapter 16 presents Gauss's tour-de-force: a proof that a heptadecagon (a regular polygon with seventeen sides) can be constructed using a straightedge and compass. By a clever argument on the symmetry of the roots of polynomials, he obtained a formula that uses only the four arithmetic operators and square roots. Gauss did not give an explicit construction of a heptadecagon, so the elegant construction by James Callagy is presented. The chapter concludes with constructions of a regular pentagon based on Gauss's method for the construction of a heptadecagon.

To keep the book as self-contained as possible, Appendix A collects proofs of theorems of geometry and trigonometry that may not be familiar to the reader.

## Style

- The reader is assumed to have a good knowledge of secondary-school mathematics, including:

  - Algebra: polynomials and division of polynomials, *monic* polynomials—those whose coefficient of the highest power is 1, quadratic equations, multiplication of expressions with exponents $a^m \cdot a^n = a^{m+n}$.
  - Euclidean geometry: congruent triangles $\triangle ABC \cong \triangle DEF$ and the criteria for congruence, similar triangles $\triangle ABC \sim \triangle DEF$ and the ratios of their sides, circles and their inscribed and central angles.
  - Analytic geometry: the cartesian plane, computing lengths and slopes of line segments, the formula for a circle.
  - Trigonometry: the functions sin, cos, tan and the conversions between them, angles in the unit circle, the trigonometric functions of angles reflected around an axis such as $\cos(180° - \theta) = -\cos\theta$.

- Statements to be proved are called *theorems* with no attempt to distinguish between theorems, lemmas and corollaries.
- When a theorem follows a construction, the variables that appear in the theorem refer to labeled points, lines and angles in the figure accompanying the construction.
- The full names of mathematicians have been given without biographical information that can be found easily in Wikipedia.
- The book is written so that it is as self-contained as possible, but occasionally the presentation depends on advanced mathematical concepts and theorems that are given without proofs. In such cases, a summary of the material is presented in boxes which may be skipped.
- There are no exercises but the ambitious reader is invited to prove each theorem before reading the proof.

- Geometric constructions can be studied using software such as Geogebra.
- $\overline{AB}$ is used both for the name of a line segment and for the length of the segment.
- $\triangle ABC$ is used both for the name a triangle and for the area of the triangle.

## Acknowledgments

This book would never have been written without the encouragement of Abraham Arcavi who welcomed me to trespass on his turf of mathematics education. He also graciously wrote the foreword. Avital Elbaum Cohen and Ronit Ben-Bassat Levy were always willing to help me (re-)learn secondary-school mathematics. Oriah Ben-Lulu introduced me to the mathematics of origami and collaborated on the proofs. I am grateful to Michael Woltermann for permission to use several sections of his reworking of Heinrich Dörrie's book. Jason Cooper, Richard Kruel, Abraham Arcavi and the anonymous reviewers provided helpful comments.

I would like to thank the team at Springer for their support and professionalism, in particular the editor Richard Kruel.

The book is published under the Open Access program and I would like to thank the Weizmann Institute of Science for funding the publication.

The LaTeX source files for the book (which include the TikZ source for the diagrams) are available at:

https://github.com/motib/surprises

*Mordechai (Moti) Ben-Ari*

# Contents

# Chapter 1
# The Collapsing Compass

A modern compass is a *fixed compass*: the distance between the two legs can be fixed so that it is possible to copy a line segment or a circle from one position to another (Fig. 1.1a). Euclid used a *collapsing compass* where a fixed distance cannot be maintained (Fig. 1.1b). Teachers often use a collapsing compass consisting of a marker tied to a string that is used to construct a circle on a whiteboard. It is impossible to maintain a fixed length when the compass is removed from the whiteboard.

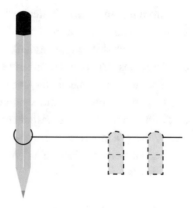

**Fig. 1.1a** A fixed compass. One leg has a needle that is placed at the center of the circle. A pencil attached to the other leg is used to draw the circle. The legs are joined by a tight hinge so that the distance between the legs (the radius of the circle) is maintained even when the compass is lifted from the paper.

**Fig. 1.1b** A collapsing compass. The user holds a piece of string at the center of the circle. The other end of the string is tied to a pencil and is used to draw the circle. When the compass is lifted from the paper, the fingers (dashed) can easily slip to a new position.

This chapter begins with a discussion of the relevance of studying construction with a straightedge and compass (Sect. 1.1). Section 1.2 compares the two types of

© The Author(s) 2022
M. Ben-Ari, *Mathematical Surprises*, https://doi.org/10.1007/978-3-031-13566-8_1

compasses in the most elementary construction: a perpendicular bisector. Section 1.3 presents Euclid's method of copying a line segment using a collapsing compass. This proves that any construction that can be done using a fixed compass can be performed using a collapsing compass. Section 1.4 shows a proof of this theorem which seems to be correct, but does not work for all configurations of lines and points. To emphasize that one must not trust diagrams, Sect. 1.5 presents a famous alleged proof that all triangles are isoceles; the proof appears to be correct but it is not because the proof is based on an incorrect diagram.

## 1.1 Construction with a Straightedge and Compass

Construction with a straightedge and compass used to be the fundamental concept taught in Euclidean geometry. Recently, it has fallen out of favor in school curricula. It is certainly true that the topic has little, if any, practical use. As we show in Sects. 2.2, 2.3, 2.4, 3.4, the Greeks knew how to perform constructions that are impossible with a straightedge and compass by using tools only slightly more advanced. Today, using numerical methods, computers can perform constructions to any desired precision.

Nevertheless, I believe that there are advantages to studying constructions:

- It is more fun and more challenging to learn geometry through constructions than simply to read theorems and proofs.
- Significant breakthroughs in mathematics have been achieved by attempts to find constructions. Chapter 16 presents a construction by Gauss that led to modern abstract algebra, in particular, the theory developed by Évariste Galois.
- It is somewhat counterintuitive and therefore very interesting that it can be proved that it is impossible to construct some geometric objects.
- Sadly, there a many people who waste years of their lives trying to perform impossible constructions. Students should certainly be aware of the futility of such efforts.

## 1.2 Fixed Compasses and Collapsing Compasses

Some geometry textbooks present the construction of a perpendicular bisector of a line segment by constructing two circles centered at the ends of the line segment such that the radii are equal and *greater than half the length of the segment* (Fig. 1.2a). This can only be done with a fixed compass because after drawing the circle centered at $A$, the distance between the legs of the compass needs to remain fixed to draw the circle centered at $B$.

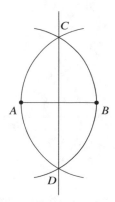

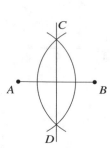

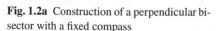

**Fig. 1.2a** Construction of a perpendicular bisector with a fixed compass

**Fig. 1.2b** Construction of a perpendicular bisector with a fixed or a collapsing compass

Figure 1.2b shows the construction of a perpendicular bisector with either a fixed
or a collapsing compass. Two circles are constructed: one centered at $A$ with radius
$\overline{AB}$ and one centered at $B$ with radius $\overline{BA}$. This can be done with a collapsing compass
because (obviously) $\overline{AB} = \overline{BA}$, so the compass does not have to "remember" the
length of $\overline{AB}$ to construct a circle centered at $B$ with the same radius. The proof
that the line constructed shown in Fig. 1.2a is a perpendicular bisector is not at all
elementary because relatively advanced concepts like congruent triangles have to
be used. However, the proof that the construction of a perpendicular bisector shown
in Fig. 1.2b is correct is simple and based on the fact that $\triangle ABC$ is an equilateral
triangle. In fact, this is the first proposition in Euclid's *Elements*. $\overline{AC} = \overline{AB}$ since they
are radii of the same circle and, similarly, $\overline{BC} = \overline{BA}$. We have: $\overline{AC} = \overline{AB} = \overline{BA} = \overline{BC}$.

Figure 1.3a shows that for the construction with a fixed compass, the triangle will
be an isosceles, not necessarily an equilateral, triangle (Fig. 1.3b).

## 1.3 Euclid's Construction for Copying a Line Segment

The second proposition of Euclid's *Elements* describes how to copy a given line
segment $\overline{AB}$ to a segment of the same length, one of whose end points is a given
point $C$. Therefore, a fixed compass adds no additional capabilities and a collapsing
compass is sufficient, although constructions are easier with a fixed compass.

**Theorem 1.1** *Given a line segment $\overline{AB}$ and a point $C$, a line segment $\overline{CC'}$, one of
whose endpoints is $C$, can be constructed using a collapsing compass, such that
$\overline{AB} = \overline{CC'}$ (Fig. 1.4a).*

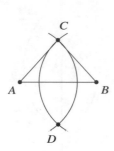

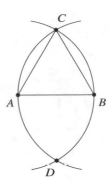

**Fig. 1.3a** Construction of an isoceles triangle with a fixed compass

**Fig. 1.3b** Construction of an equilateral triangle with a collapsing compass

***Proof*** Construct the line segment $\overline{AC}$. Construct the equilateral triangle $\triangle ACD$ whose base is $\overline{AC}$ (Fig. 1.4b). By Euclid's first proposition, the triangle can be constructed using a collapsing compass. Construct the ray that is an extension of the line segment *from D to A*, and construct the ray that is an extension of the line segment *from D to C* (Fig. 1.5a). Construct the circle centered at $A$ with radius $\overline{AB}$ and denote the intersection of the circle and the ray extending $\overline{DA}$ by $E$ (Fig. 1.5b). Construct the circle centered at $D$ with radius $\overline{DE}$ and denote the intersection of the circle and the ray extending $\overline{DC}$ by $F$ (Fig. 1.6).

$\overline{DC} = \overline{DA}$ because $\triangle ACD$ is equilateral. $\overline{AE} = \overline{AB}$ are radii of the same circle, as are $\overline{DF} = \overline{DE}$. Therefore:

$$\overline{CF} = \overline{DF} - \overline{DC} = \overline{DE} - \overline{DC} = \overline{DE} - \overline{DA} = \overline{AE} = \overline{AB}. \qquad \square$$

The specification of the directions of the rays is essential. The proof here works for any line segment $\overline{AB}$ and any point $C$, regardless of its position relative to $\overline{AB}$. By specifying directions the "cone" enclosed by the two rays will intersect the circles even if $\overline{AC} > \overline{AB}$ (Fig. 1.7).

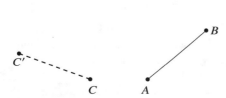

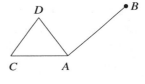

**Fig. 1.4a** Copy the line segment $\overline{AB}$. The orientation of $\overline{CC'}$ is not important.

**Fig. 1.4b** Copying a line segment with a collapsing compass

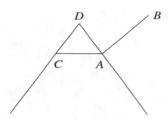

**Fig. 1.5a** Constructing rays from $D$

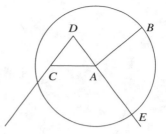

**Fig. 1.5b** Constructing a circle with radius $\overline{AB}$

## 1.4 A Flawed Construction for Copying a Line Segment

***Proof*** Construct three circles: one centered at $A$ with radius $\overline{AB}$, one centered at $A$ with radius $\overline{AC}$, and one centered at $C$ with radius $\overline{AC} = \overline{CA}$. Denote the intersections of the circles centered at $A$ and $C$ by $E$ and $F$, respectively, and denote an intersection of the circle centered at $C$ and the circle centered at $A$ with radius $\overline{AB}$ by $D$. If $\overline{AC} > \overline{AB}$, the construction is as shown in Fig. 1.8.

Construct a circle centered at $E$ with radius $\overline{ED}$. Denote the intersection of this circle with the circle centered at $A$ with radius $\overline{AC}$ by $G$. There are two intersections, so choose the one closer to $C$ (Fig. 1.9). $\overline{CD} = \overline{CE}$ are radii of the same circle as are $\overline{AE} = \overline{AG}$. By construction the radii $\overline{CE}$ and $\overline{AE}$ are equal. Therefore,

$$\overline{CD} = \overline{CE} = \overline{AE} = \overline{AG}.$$

$\overline{EG} = \overline{ED}$ are radii of the same circle, so $\triangle EAG \cong \triangle DCE$ by side-side-side and $\angle GEA = \angle DEC$.

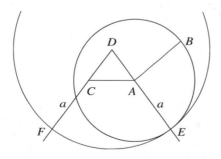

**Fig. 1.6** Construction of $\overline{CF} = \overline{AB}$

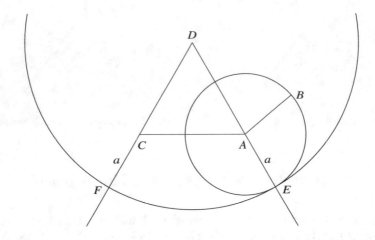

**Fig. 1.7** Construction for $\overline{AC} > \overline{AB}$

Since:

$$\angle GEC = \angle GEA - \angle CEA = \angle DEC - \angle CEA = \angle DEA\,,$$

it follows that $\triangle ADE \cong \triangle CGE$ by side-angle-side. $\overline{AB} = \overline{AD}$ are radii of the smaller circle centered at $A$, so $\overline{GC} = \overline{AD} = \overline{AB}$.                                      □

The proof is correct only if $\overline{AC} > \overline{AB}$. Figure 1.10 shows a diagram where $\overline{AC} < \overline{AB}$ and you can see that $\overline{AB} \neq \overline{GC}$.

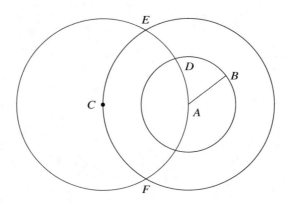

**Fig. 1.8** Construction for copying a line segment (1)

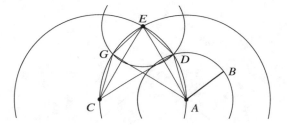

**Fig. 1.9** Construction for copying a line segment (2)

## 1.5 Don't Trust a Diagram

**Theorem 1.2 (Incorrect, of course)** *All triangles are isosceles.*

***Proof (Incorrect)*** Given an arbitrary triangle $\triangle ABC$, let $P$ be the intersection of the angle bisector of $\angle BAC$ and the perpendicular bisector of $\overline{BC}$. The intersections of the altitudes from $P$ to the sides $\overline{AB}, \overline{AC}$ are denoted by $E, F$, respectively (Fig. 1.11). $\triangle APE \cong \triangle APF$ because they are right triangles with equal angles $\alpha$ and common side $\overline{AP}$. $\triangle DPB \cong \triangle DPC$ since they are right triangles, $\overline{PD}$ is a common side and $\overline{BD} = \overline{CD} = a$. $\triangle EPB \cong \triangle FPC$ since they are right triangles, $\overline{EP} = \overline{PF}$ by the first congruence and $\overline{PB} = \overline{PC}$ by the second congruence. By combining the equations we get that $\triangle ABC$ is isoceles:

$$\overline{AB} = \overline{AE} + \overline{EB} = \overline{AF} + \overline{FC} = \overline{AC}.$$

□

The *logic* of the proof is correct, but the diagram upon which the proof is based is not correct because point $P$ is *outside* the triangle (Fig. 1.12).

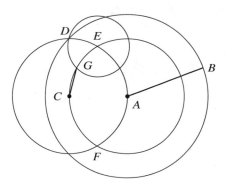

**Fig. 1.10** A diagram for which the proof doesn't work

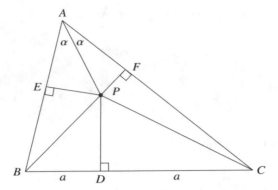

**Fig. 1.11** An incorrect proof that all triangles are isoceles

## What Is the Surprise?

As a student I took it for granted that a compass has a friction joint that maintains the distance between the point and the pencil when it is lifted from the paper. When the teacher used a compass made from a piece of string and a piece of chalk, I never imagined that it differed from my compass. The article by Gotfried Toussaint was a real surprise, as was his demonstration that post-Euclid proofs were incorrect because they depended on diagrams that made unwarranted assumptions. I recommend the article to readers who wish to deepen their understanding of proofs in mathematics.

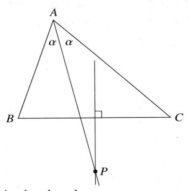

**Fig. 1.12** Why the construction doesn't work

## Sources

This chapter is based on [50]. The incorrect construction of the equivalence of the two compasses in Sect. 1.4 is from [37]. A comprehensive English translation of Euclid's *Elements* together with an extensive commentary [22] was written by Thomas L. Heath, one of the foremost experts in Greek mathematics.

# Chapter 2
# Trisection of an Angle

It is impossible to trisect an arbitrary angle (divide the angle into three equal parts) using only a straightedge and compass. Trisection requires the construction of cube roots, but a straightedge and compass can only construct lengths that are expressions built from integers, the four arithmetic operators and square roots. This was proved by Pierre Wantzel in 1837. Nevertheless, innumerable amateurs continue to attempt to trisect an angle. Their constructions are approximations though they are convinced that the constructions are correct. Section 2.1 presents two such constructions, develops formulas for the angles and shows the errors in the approximations.

Greek mathematicians discovered that if other instruments are allowed, angles can be trisected. Section 2.2 explains a construction by Archimedes using a simple instrument called a *neusis* and Sect 2.3 shows how to double a cube using the neusis. Section 2.4 presents a construction for trisection by Hippias using an instrument called a *quadratrix*. The rest of the chapter contains a proof of the impossibility of trisecting an angle. Section 2.5 characterizes constructible numbers, Sect. 2.6 relates constructible numbers to roots of polynomials and Sect. 2.7 uses this theory to show that trisecting an angle and doubling a cube are impossible.

## 2.1 Approximate Trisections

### 2.1.1 First Approximate Trisection

**Construction:** Let $\theta = \angle AOB$ be an arbitrary angle and without loss of generality assume that $A, B$ are on a unit circle whose center is $O$. Bisect $\angle AOB$ and let $C$ be the intersection of the bisector with the unit circle. Let $D$ be the midpoint of $\overline{OA}$ and let $T$ be the midpoint of the $\overline{DC}$. Denote the angle $\angle DOT$ by $\phi$ (Fig. 2.1).

© The Author(s) 2022
M. Ben-Ari, *Mathematical Surprises*, https://doi.org/10.1007/978-3-031-13566-8_2

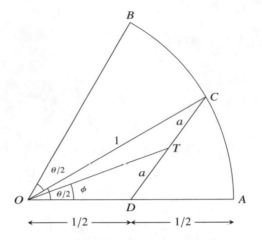

**Fig. 2.1** First approximate trisection (1)

**Theorem 2.1**

$$\tan \phi = \frac{2\sin(\theta/2)}{1 + 2\cos(\theta/2)}.$$

*Proof* Figure 2.2 is extracted from Fig. 2.1 and contains additional annotations.

Let $\overline{CF}$ be the perpendicular to $\overline{OA}$ that intersects $\overline{OA}$ at $F$. Since $\overline{OC} = 1$, $\overline{CF} = \sin(\theta/2)$ and $\overline{OF} = \cos(\theta/2)$. Let $\overline{TE}$ be the perpendicular to $\overline{OA}$ that intersects $\overline{OA}$ at $E$.

$T$ is the midpoint of $\overline{DC}$ so $\overline{DT} = \overline{TC} = a$. But $\overline{FT}$ is the median to the hypotenuse of a right triangle, so $\overline{FT} = a$ and therefore $\triangle DTF$ is isoceles. It follows that $\overline{TE}$ is the both the median and the altitude of $\overline{DF}$. From the diagram it is easy to see that:

$$\overline{OE} = \frac{1}{2} + \frac{1}{2}\left(\cos\frac{\theta}{2} - \frac{1}{2}\right).$$

Compute the length $2a = \overline{CD}$ using Pythagoras's Theorem in $\triangle DCF$:

$$(2a)^2 = \left(\cos\frac{\theta}{2} - \frac{1}{2}\right)^2 + \sin^2\frac{\theta}{2}.$$

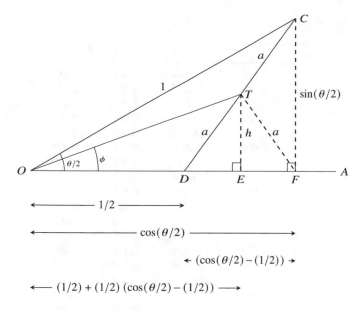

**Fig. 2.2** First approximate trisection (2)

The length $h = \overline{TE}$ can be computed from Pythagoras's Theorem in $\triangle DTE$:

$$a^2 = h^2 + \left[\frac{1}{2}\left(\cos\frac{\theta}{2} - \frac{1}{2}\right)\right]^2$$

$$h^2 = \frac{1}{4}\left(\cos\frac{\theta}{2} - \frac{1}{2}\right)^2 + \frac{1}{4}\sin^2\frac{\theta}{2} - \left[\frac{1}{2}\left(\cos\frac{\theta}{2} - \frac{1}{2}\right)\right]^2 = \frac{1}{4}\sin^2\frac{\theta}{2}$$

$$h = \frac{1}{2}\sin\frac{\theta}{2}$$

$$\tan\phi = \frac{h}{\overline{OE}} = \frac{\frac{1}{2}\sin\frac{\theta}{2}}{\frac{1}{2} + \frac{1}{2}\left(\cos\frac{\theta}{2} - \frac{1}{2}\right)} = \frac{2\sin\frac{\theta}{2}}{1 + 2\cos\frac{\theta}{2}} . \qquad \square$$

This is an approximation to a trisection $\phi = \theta/3$. For $\theta = 60°$:

$$\tan^{-1}\left(\frac{2\sin 30°}{1 + 2\cos 30°}\right) = \tan^{-1} 0.366 \approx 20.1° \approx 20° .$$

Table 2.1 shows the errors for a range of acute angles. The error is relatively small for small angles, rising to 1% at 85°.

**Table 2.1** Errors in the first approximate trisection

| $\theta(°)$ | $\theta/3(°)$ | $\tan^{-1}\phi(°)$ | Error(°) | Error(%) |
|---|---|---|---|---|
| 5 | 1.667 | 1.667 | 0.000 | 0.004 |
| 10 | 3.333 | 3.334 | 0.000 | 0.014 |
| 15 | 5.000 | 5.002 | 0.002 | 0.032 |
| 20 | 6.667 | 6.670 | 0.004 | 0.057 |
| 25 | 8.333 | 8.341 | 0.007 | 0.088 |
| 30 | 10.000 | 10.013 | 0.013 | 0.128 |
| 35 | 11.667 | 11.687 | 0.020 | 0.174 |
| 40 | 13.333 | 13.364 | 0.030 | 0.228 |
| 45 | 15.000 | 15.043 | 0.043 | 0.289 |
| 50 | 16.667 | 16.726 | 0.060 | 0.358 |
| 55 | 18.333 | 18.413 | 0.080 | 0.435 |
| 60 | 20.000 | 20.104 | 0.104 | 0.520 |
| 65 | 21.667 | 21.799 | 0.133 | 0.612 |
| 70 | 23.333 | 23.500 | 0.166 | 0.713 |
| 75 | 25.000 | 25.206 | 0.206 | 0.823 |
| 80 | 26.667 | 26.918 | 0.251 | 0.941 |
| 85 | 28.333 | 28.636 | 0.303 | 1.068 |

## 2.1.2 Second Approximate Trisection

**Construction:** Let $\theta = \angle AOB$ be an arbitrary angle and without loss of generality assume that $A, B$ are on a unit circle whose center is $O$. Construct a circle of radius $1/3$ with center $O$ and let $D$ be its intersection with $\overline{OA}$. Bisect $\angle AOB$ and let $C$ be the intersection of the bisector with the circle of radius $1/3$. Construct the chord $\overline{CD}$ and the chords $\overline{AE} = \overline{ET} = \overline{CD}$. Since equal chords subtend equal central angles $\angle TOE = \angle EOA = \phi$ (Fig. 2.3).

**Theorem 2.2**

$$\cos\phi = 1 - \frac{1}{9}(1 - \cos(\theta/2)) = 1 - \frac{2}{9}\sin^2(\theta/4).$$

***Proof*** By the Law of Cosines in $\triangle DOC$:

$$\overline{CD} = \left(\frac{1}{3}\right)^2 + \left(\frac{1}{3}\right)^2 - 2\left(\frac{1}{3}\right)\left(\frac{1}{3}\right)\cos(\theta/2) = \frac{2}{9}(1 - \cos(\theta/2)).$$

By the Law of Cosines in $\triangle EOA$:

$$\overline{AE} = 1^2 + 1^2 - 2 \cdot 1 \cdot 1 \cdot \cos\phi = 2(1 - \cos\phi).$$

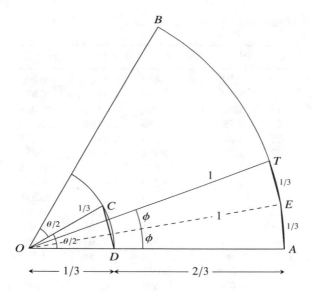

**Fig. 2.3** Second approximate trisection

Equating the two expressions for $\overline{CD} = \overline{AE}$ and simplifying we get:

$$\cos\phi = 1 - \frac{1}{9}(1 - \cos(\theta/2)).$$

Since $\cos 2\alpha = \cos^2\alpha - \sin^2\alpha = 1 - 2\sin^2\alpha$, and therefore $1 - \cos 2\alpha = 2\sin^2\alpha$, we have the alternate formula:

$$\cos\phi = 1 - \frac{2}{9}\sin^2(\theta/4).$$

□

This is an approximation to a trisection $2\phi = \theta/3$. For $\theta = 60°$:

$$2\cos^{-1}\left(1 - \frac{1}{9}(1 - \cos 30°)\right) \approx 19.8° \approx 20°.$$

Table 2.2 shows the errors for a range of acute angles. This construction is much less accurate than the one in Sect. 2.1.1.

**Table 2.2** Errors in the second approximate trisection

| $\theta(°)$ | $\theta/3(°)$ | $\cos^{-1}2\phi(°)$ | Error$(°)$ | Error$(\%)$ |
|---|---|---|---|---|
| 5 | 1.667 | 1.667 | 0.000 | 0.007 |
| 10 | 3.333 | 3.332 | 0.001 | 0.028 |
| 15 | 5.000 | 4.997 | 0.003 | 0.063 |
| 20 | 6.667 | 6.659 | 0.008 | 0.113 |
| 25 | 8.333 | 8.319 | 0.015 | 0.176 |
| 30 | 10.000 | 9.975 | 0.025 | 0.254 |
| 35 | 11.667 | 11.626 | 0.040 | 0.346 |
| 40 | 13.333 | 13.273 | 0.060 | 0.451 |
| 45 | 15.000 | 14.914 | 0.086 | 0.571 |
| 50 | 16.667 | 16.549 | 0.118 | 0.705 |
| 55 | 18.333 | 18.177 | 0.156 | 0.853 |
| 60 | 20.000 | 19.797 | 0.203 | 1.015 |
| 65 | 21.667 | 21.408 | 0.258 | 1.192 |
| 70 | 23.333 | 23.011 | 0.322 | 1.382 |
| 75 | 25.000 | 24.603 | 0.397 | 1.586 |
| 80 | 26.667 | 26.185 | 0.481 | 1.805 |
| 85 | 28.333 | 27.756 | 0.577 | 2.038 |

## 2.2 Trisection Using a Neusis

The term *straightedge* is used instead of *ruler* because a straightedge has no marks on it. It can only be used to construct a straight line between two given points. Archimedes showed that a *neusis*, a straightedge with two marks that are a fixed distance apart, can be used to trisect an angle (Fig. 2.4). We define the distance between the marks to be 1.

**Construction:** Let $\alpha = \angle ABE$ be an arbitrary angle in a unit circle with center $B$, where the radius of the circle equals the distance between the marks on the neusis. Extend the radius $\overline{EB}$ beyond the circle. Place an edge of the neusis on $A$ and move it until it intersects the extension of $\overline{EB}$ at $D$ and the circle at $C$, using the marks so that the length of the line segment $\overline{CD}$ is 1.[1] Construct the line $\overline{AD}$. Denote $\angle CDB = \beta$ (Fig. 2.5).

**Fig. 2.4** A neusis

---

[1] This operation is called *verging*.

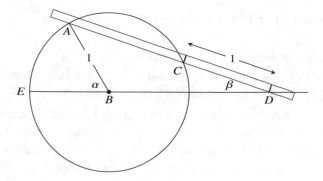

**Fig. 2.5** The neusis construction for trisecting an angle (1)

**Theorem 2.3** $\beta = \alpha/3$.

**Proof** Construct $\overline{BC}$ and denote the angles and line segments as shown in Fig. 2.6. $\triangle ABC$ and $\triangle BCD$ are isoceles triangles: $\overline{AB} = \overline{BC}$ are radii of the same circle and $\overline{BC} = \overline{CD}$ by construction using the neusis. Since the sum of the angles of a triangle is equal to 180° and the sum of supplementary angles is also equal to 180°, we have:

$$\epsilon = 180° - 2\beta$$
$$\gamma = 180° - \epsilon = 2\beta$$
$$\delta = 180° - 2\gamma = 180° - 4\beta$$
$$\alpha = 180° - \delta - \beta = 3\beta.$$                                □

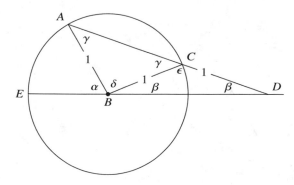

**Fig. 2.6** The neusis construction for trisecting an angle (2)

## 2.3 Doubling the Cube with a Neusis

Given a cube $C$ construct another cube with twice its volume. If the volume of $C$ is $V$ its sides are of length $\sqrt[3]{V}$. The sides of a cube with twice the volume are $\sqrt[3]{2V} = \sqrt[3]{2} \cdot \sqrt[3]{V}$, so if we can construct $\sqrt[3]{2}$ we can double the cube.

**Construction:** Construct the unit equilateral triangle $\triangle ABC$ and extend $\overline{CA}$ with another unit line segment to $D$. Construct rays extending $\overline{AB}$ and $\overline{DB}$. Place the neusis on point $C$ and move it until one mark on the neusis is placed on the ray $\overline{AB}$ at $P$ and the other mark is placed on the ray $\overline{DB}$ at $Q$. Denote $\overline{CQ} = x$ and $\overline{BP} = y$ (Fig. 2.7).

**Theorem 2.4** $x = \sqrt[3]{2}$.

**Proof** Since $\triangle ABC$ is equilateral, $\cos \angle CAP = \cos 60° = \frac{1}{2}$ and by the Law of Cosines in $\triangle APC$:

$$\overline{CP} = \overline{AC}^2 + \overline{AP}^2 - 2 \cdot \overline{AC} \cdot \overline{AP} \cos 60° \tag{2.1a}$$

$$(x+1)^2 = 1^2 + (y+1)^2 - 2 \cdot 1 \cdot (y+1) \cdot \frac{1}{2} \tag{2.1b}$$

$$x^2 + 2x = y^2 + y . \tag{2.1c}$$

By Menelaus's theorem (Thm. A.20):

$$\frac{\overline{AB}}{\overline{BP}} \cdot \frac{\overline{PQ}}{\overline{QC}} \cdot \frac{\overline{CD}}{\overline{DA}} = 1 .$$

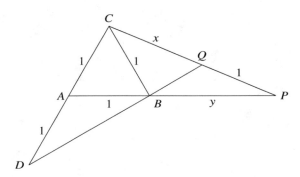

**Fig. 2.7** Doubing the cube with a neusis

Therefore:

$$\frac{1}{y} \cdot \frac{1}{x} \cdot \frac{2}{1} = 1 \tag{2.2a}$$

$$xy = 2. \tag{2.2b}$$

Substituting Eq. 2.2b into Eq. 2.1c gives:

$$x^2 + 2x = \frac{4}{x^2} + \frac{2}{x}$$
$$x^4 + 2x^3 = 4 + 2x$$
$$x^3(x + 2) = 2(x + 2)$$
$$x = \sqrt[3]{2}. \qquad\qquad \Box$$

## 2.4  Trisection Using a Quadratrix

Let $\overline{ABCD}$ be a square. Let $l_1$ be a line segment placed initially at $\overline{DC}$ and let $l_2$ be a line segment placed initially at $\overline{AD}$. Move $l_1$ move at a constant linear velocity until it reaches $\overline{AB}$ and rotate $l_2$ at a constant angular velocity clockwise on $A$ until it also reaches $\overline{AB}$. Assume that they reach $\overline{AB}$ together. For example, if $l_2$ rotates at $1°$/second and the side of the square is 9 centimeters, $l_1$ must move at 0.1 cm/second. The trace of their point of intersection $P$ is called a *quadratrix curve* or simply a *quadratrix* (Fig. 2.8a). Its definition is attributed to the mathematician Hippias.

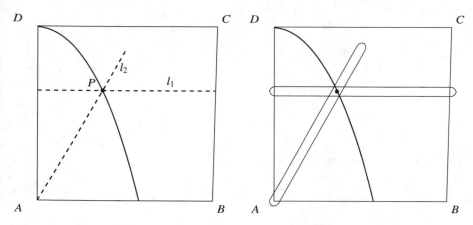

**Fig. 2.8a**  A quadratrix curve                 **Fig. 2.8b**  A quadratrix compass

A quadratrix can be constructed using a *quadratrix compass* as shown in Fig. 2.8b. It consists of two (unmarked) straightedges that move as described above. A joint constrains them to move together and traces out the curve.

A quadratrix can be used to trisect an angle.

**Construction:** Let $\angle CDP_1 = \alpha$ be an arbitrary angle, where $P_1$ is the intersection of the line defining the angle $\alpha$ relative to $\overline{DC}$ and the quadratrix. Construct a line through $P_1$ parallel to $\overline{DC}$ and denote its intersection with $\overline{AD}$ by $E$. Denote the line segment $\overline{DE}$ by $t$ and trisect it (Sect. 2.5) to obtain point $F$ that is $t/3$ from $\overline{DC}$. Let $P_2$ be the intersection of a line from $F$ parallel to $\overline{DC}$ and the quadratrix, and denote by $\theta$ the angle between $\overline{DC}$ and $\overline{DP_2}$ (Fig. 2.9).

**Theorem 2.5** $\theta = \alpha/3$.

**_Proof_** $E$ has $y$-coordinate $1 - t$ so by construction $F$ has $y$-coordinate $1 - (t/3)$. Since the constant linear velocity of the horizontal line is proportional to the constant angular velocity of the rotating line $\theta/\alpha = (t/3)/t$ and $\theta = \alpha/3$.                     □

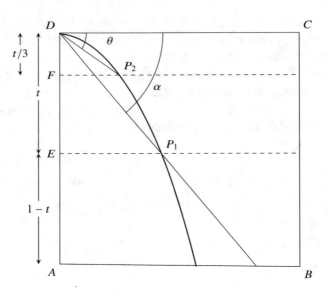

**Fig. 2.9** Trisection of an angle using a quadratrix

## 2.5 Constructible Numbers

Let $l$ be a line segment defined to be of length 1.

**Definition 2.1** A number $a$ is *constructible* if and only if a line segment of length $a$ can be constructed with a straightedge and compass starting from $l$.

Given line segment $l = \overline{AB}$, construct a line containing $\overline{AB}$ and use the compass to find a point $C$ on the line that is a distance of 1 from $B$. Then $\overline{AC}$ is of length 2 so the number 2 is constructible. A line segment $\overline{BD}$ of length 1 can be constructed perpendicular to $\overline{AB}$ at $B$. The hypotenuse of the triangle $\triangle ABD$ is of length $\sqrt{2}$ so the number $\sqrt{2}$ is constructible.

**Theorem 2.6** *A number is* constructible *if and only if it is the value of an expression built from the integers, the four arithmetic operations* $\{+, -, \times, /\}$ *and the operation of taking a square root* $\sqrt{}$.

***Proof*** First we show that the values of these expressions are constructible.

**Addition and subtraction:** Given line segments $\overline{PQ} = a$ and $\overline{RS} = b$, construct a circle centered at $Q$ with radius $b$ (Fig. 2.10). Extend $\overline{PQ}$ until it intersects the circle at $U$. Then $\overline{PTQU}$ is a line segment, where $\overline{PT} = a - b$ and $\overline{PU} = a + b$.

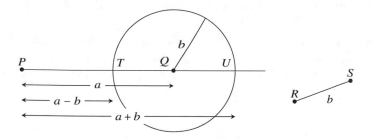

**Fig. 2.10** Construction of addition and subtraction

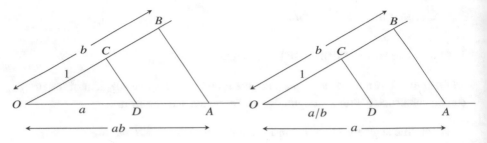

**Fig. 2.11a** Construction of multiplication          **Fig. 2.11b** Construction of division

**Multiplication:** By similar triangles in Fig. 2.11a, $(1/b) = (a/\overline{OA})$, so $\overline{OA} = ab$.

**Division:** By similar triangles in Fig. 2.11b, $(1/b) = (\overline{OD}/a)$, so $\overline{OD} = (a/b)$.

**Square roots:** Given a line segment $\overline{BC} = a$, construct $\overline{AB} = 1 + a$ and a semicircle with $\overline{AB}$ as its diameter. Construct a perpendicular at $C$ and let $D$ be the intersection of the perpendicular and the circle (Fig. 2.12). $\angle ADB$ is a right angle because it is subtended by a diameter. By similar triangles $(h/1) = (a/h)$, so $h^2 = a$ and $h = \sqrt{a}$.

To prove the converse of the theorem, we need to determine what expressions can be constructed by a straightedge and compass. There are three constructions:[2]

1. Two lines intersect in a point (Fig. 2.13a). The coordinates of the intersection can be derived from the equations of the two lines $y = x$ and $y = 4x - 2$. The point of intersection is $P = (2/3, 2/3)$.
2. A line intersects a circle in zero, one or two points (Fig. 2.13b). The coordinates of the intersections can be derived from the equations of the line $y = x$ and the circle $x^2 + y^2 = 4$. The points of intersection are $P = (\sqrt{2}, \sqrt{2})$ and $Q = (-\sqrt{2}, -\sqrt{2})$.

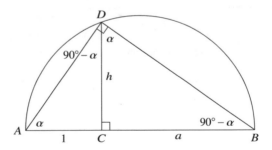

**Fig. 2.12** Construction of a square root

---

[2] For clarity these are illustrated for specific values rather than the most general equations.

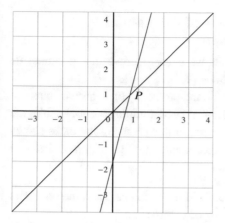
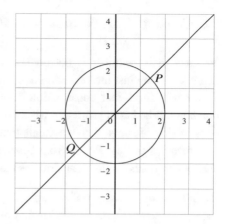

**Fig. 2.13a** The point of intersection of two lines

**Fig. 2.13b** The points of intersection of a line and a circle

3. Two circles intersect in zero, one or two points (Fig. 2.14). The coordinates of the intersections can be derived from the equations of the two circles $(x-1)^2+y^2 = 4$, $(x+1)^2 + y^2 = 4$. The points of intersection are $P = (0, \sqrt{2})$, $Q = (0, -\sqrt{2})$. ☐

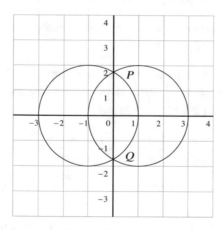

**Fig. 2.14** The points of intersection of two circles

## 2.6 Constructible Numbers As Roots of Polynomials

To show that a number is not constructible, we need to prove that it cannot be expressed using just integers and the operations $\{+, -, \times, /, \sqrt{}\}$.

We will show that constructible numbers are the roots of a certain class of polynomials and then prove that trisecting an angle and doubling a cube require the construction of roots of polynomials that are not in this class. Today these results are proved using field theory from abstract algebra, but here I give a proof that uses elementary mathematics. The proof is based on the following definition.

**Definition 2.2** The *depth* of an expression built from the integers and the operators $\{+, -, \times, /, \sqrt{}\}$ is the maximum level of nesting of square roots.

*Example 2.1* Consider the following expression:

$$\sqrt{17 + 3\sqrt{17} - \sqrt{34 - 2\sqrt{17}} - 2\sqrt{34 + 2\sqrt{17}}}.$$

The depth is 3 because at the right of the expression we have $\sqrt{17}$ which is nested within $\sqrt{34 + 2\sqrt{17}}$, which in turn is nested within $\sqrt{17 + \cdots - \cdots - 2\sqrt{34 + 2\sqrt{17}}}$.

**Theorem 2.7** *A expression of depth $n$ can be expressed as $a + b\sqrt{c}$ where $a, b, c$ are expressions of depth at most $n - 1$.*

**Proof** Simple computations show that the expressions $(a_1 + b_1\sqrt{c}) \, op \, (a_2 + b_2\sqrt{c})$ for the operators $op = \{+, -, \times\}$ result in expressions $a + b\sqrt{c}$ of depth $n - 1$. For division the computation is a bit more complicated:

$$\frac{a_1 + b_1\sqrt{c}}{a_2 + b_2\sqrt{c}} = \frac{(a_1 + b_1\sqrt{c})(a_2 - b_2\sqrt{c_2})}{(a_2 + b_2\sqrt{c})(a_2 - b_2\sqrt{c})}$$
$$= \frac{a_1 a_2 - b_1 b_2 c}{a_2^2 - b_2^2 c} + \frac{a_2 b_1 - a_1 b_2}{a_2^2 - b_2^2 c}\sqrt{c},$$

which is of the form $a + b\sqrt{c}$ and of depth $n - 1$. Finally, the square root of an expression of depth $n - 1$ is an expression of depth $n$.                                     $\square$

**Theorem 2.8** *Let $p(x)$ be a monic cubic polynomial with rational coefficients:*

$$p(x) = x^3 + a_2 x^2 + a_1 x + a_0,$$

*and let $r = a + b\sqrt{c}$ be a root of $p(x)$ of minimal depth $n$, where $a, b, c$ are of depth (at most) $n - 1$. Then $r' = a - b\sqrt{c}$ is a root of $p(x)$ and $r \neq r'$.*

**Proof** Let us compute $p(r)$ which is equal to 0 since $r$ is a root:

$$(a + b\sqrt{c})^3 + a_2(a + b\sqrt{c})^2 + a_1(a + b\sqrt{c}) + a_0 =$$
$$(a^3 + 3a^2b\sqrt{c} + 3ab^2c + b^3c\sqrt{c})$$
$$+ a_2(a^2 + 2ab\sqrt{c} + b^2c) + a_1(a + b\sqrt{c}) + a_0 =$$
$$(a^3 + 3ab^2c + a_2a^2 + a_2b^2c + a_1a + a_0)$$
$$+ (3a^2b + b^3c + 2a_2ab + a_1b)\sqrt{c} =$$
$$d + e\sqrt{c} = 0.$$

where $d, e$ are expressions of depth $n - 1$ formed from the rational coefficients and $a, b, c$. Then $\sqrt{c} = -d/e$, so $a + b\sqrt{c}$ can be expressed as an expression of depth $n - 1$, contracting the assumption that $a + b\sqrt{c}$ is of minimal depth $n$. Since $\sqrt{c} \neq 0$ and is of depth $n$, for $d + e\sqrt{c}$ to be zero it must be that $d = e = 0$.

Consider now $r' = a - b\sqrt{c}$. By examining the above computation we see that $p(r') = d - e\sqrt{c} = 0 + 0 \cdot \sqrt{c} = 0$, so $r'$ is also a root of $p$.

If $r = r'$ then $0 = r - r' = 2b\sqrt{c}$, which is true only if $b = 0$ so $r, r'$ would be of depth $n - 1$, again contradicting the assumption. $\qquad\square$

**Theorem 2.9** *If a monic cubic polynomial with rational coefficients:*

$$p(x) = x^3 + a_2x^2 + a_1x + a_0$$

*has no rational roots then none of its roots is constructible.*

**Proof** By the Fundamental Theorem of Algebra (Thm. 16.1) $p(x)$ has three roots $r_1, r_2, r_3$. Let $r_1 = a + b\sqrt{c}$ be a root of minimal depth $n$. By the assumption that there are no rational roots, $n \geq 1$, and therefore $b \neq 0$ and $c \neq 0$. By Thm. 2.8, $r_2 = a - b\sqrt{c}$ is also a root. Perform the following multiplication:

$$(x - r_1)(x - r_2)(x - r_3) = x^3 - (r_1 + r_2 + r_3)x^2 \qquad (2.3a)$$
$$+ (r_1r_2 + r_1r_3 + r_2r_3)x + r_1r_2r_3 \qquad (2.3b)$$
$$a_2 = -(r_1 + r_2 + r_3) \qquad (2.3c)$$
$$r_3 = -(a_2 + r_1 + r_2). \qquad (2.3d)$$

Since $a_2$ is rational so is:

$$r_3 = -a_2 - (r_1 + r_2) = -a_2 - 2a,$$

contradicting the assumption. $\qquad\square$

## 2.7  Impossibility of the Classical Constructions

**Theorem 2.10** $\sqrt[3]{2}$ *is irrational.*

**Proof** Assume that $\sqrt[3]{2}$ is rational and equal to $p/q$ where $p, q$ are integers with no common factors other than $\pm 1$. Then:

$$(p/q)^3 = (\sqrt[3]{2})^3$$
$$p^3 = 2q^3 ,$$

so $p$ must be divisible by 2, say $p = 2r$. Now:

$$8r^3 = 2q^3$$
$$q^3 = 4r^3 ,$$

so $q$ is divisible by 2, contradicting the assumption that $p, q$ have no common factor.□

**Theorem 2.11** $x^3 - 2$ *has no rational roots so it is impossible to double a cube with a straightedge and compass.*

**Proof** One of its roots is $\sqrt[3]{2}$ which by Thm. 2.10 is irrational. The other roots are the roots of the quadratic equation $x^2 + \sqrt[3]{2}x + (\sqrt[3]{2})^2$ obtained by dividing $x^3 - 2$ by $x - \sqrt[3]{2}$. It is easy to check that its roots are not rational (in fact, not even real).     □

**Theorem 2.12** *It is impossible to trisect an arbitrary angle with a straightedge and compass.*

**Proof** It is sufficient to show the impossibility for one angle. Let us try to trisect $60°$ to obtain $20°$. By Thm. A.6:

$$\cos 3\alpha = 4\cos^3 \alpha - 3\cos \alpha$$
$$\cos 60° = 4\cos^3 20° - 3\cos 20° .$$

Denote $x = \cos 20°$ and $2x$ by $y$. Since $\cos 60° = 1/2$ we have:

$$4x^3 - 3x - \frac{1}{2} = 0$$
$$8x^3 - 6x - 1 = 0$$
$$y^3 - 3y - 1 = 0 .$$

To prove that the polynomial $y^3 - 3y - 1$ has no rational roots suppose that $y = a/b$ is a rational root with $a, b$ having no common factor other than $\pm 1$. Then:

$$(a/b)^3 - 3(a/b) - 1 = 0 \tag{2.4a}$$

$$a^3 - 3ab^2 = b^3 \tag{2.4b}$$

$$a(a - 3b^2) = b^3 \tag{2.4c}$$

$$a^3 = b(b^2 + 3ab). \tag{2.4d}$$

By Eq. 2.4c, $b$ must be divisible by $a$, and by Eq. 2.4d, $a$ must be divisible by $b$, which is possible only if $a = b = \pm 1$ and $a/b = \pm 1$. By computation, $y = a/b = 1$ and $y = a/b = -1$ are not roots of the polynomial. $\square$

An alternate way of proving the impossibility of the constructions is to use the following theorem which we present without proof.

**Theorem 2.13** *If a monic polynomial $p(x) = x^n + a_{n-1}x^{n-1} + \cdots + a_0$ with integer coefficients has rational roots then it has integer roots.*

To show the impossibility of duplicating a cube we need to show that:

$$x^3 - 2 = (x - r_2)(x - r_1)(x - r_0)$$

has no integer roots. Since $r_0 r_1 r_2 = -2$, all roots must divide 2, so the only possible integer roots are $\pm 1, \pm 2$. A quick computation shows that none of them are roots.

To show the impossibility of trisecting an angle we need to show that $y^3 - 3y - 1$ has no integer roots. An integer root must divide $-1$ but neither 1 nor $-1$ are roots.

## What Is the Surprise?

Underwood Dudley has made an extensive study of what he calls "cranks" who waste years of their lives trying to trisect angles with a straightedge and compass. Not only do they delude themselves into thinking that this is possible, but, even worse, they think that a solution would be important. Of course, a solution would have no practical use, since tools such as the neusis and quadratrix can solve the problem exactly. The sheer number of such constructions is surprising, especially since many of them are clever and achieve good approximations. Computing the formulas associated with the constructions is an excellent exercise in trigonometry.

It is also surprising that proofs of the impossibility of these geometric constructions are purely algebraic using properties of roots of polynomials.

## Sources

Wikipedia [51, 58, 62] is a good source for the constructions in this chapter. The two approximate trisections are from [15, pp. 67–68, 95–96]. The second example is attributed to the famous philosopher Thomas Hobbes. Both [31, pp. 48–49] and [15, pp. 6–7] discuss trisection using the quadratrix. The doubling of the cube using a neusis is taken from [14].

A rigorous treatment of constructibility can be found in textbooks on abstract algebra such as [17], which contains a general proof of the converse of Thm 2.6 in Sect. 32. Theorem 2.13 is Thm. 23.11 of [17]. A relatively accessible presentation of Wantzel's proof can be found in [48]. My presentation of constructibility is based upon the presentations in [11, Chap. III] and [27].

# Chapter 3
# Squaring the Circle

Squaring the circle, the construction of a square with the same area as a given circle, is one of the three construction problems that the Greeks posed but were unable to solve. Unlike trisecting the angle and doubling the cube, where the impossibility follows from properties of the roots of polynomials, the impossibility of squaring the circle follows from the transcendentality of $\pi$: it is not the root of any polynomial with rational coefficients. This is a difficult theorem that was proved in 1882 by Carl von Lindemann.

Approximations to $\pi \approx 3.14159265359$ have been known since ancient times. Some simple but reasonably accurate approximations are:

$$\frac{22}{7} \approx 3.142857, \quad \frac{333}{106} \approx 3.141509, \quad \frac{355}{113} \approx 3.141593.$$

We present three constructions by a straightedge and compass of approximations to $\pi$. One is by by Adam Kochański (Sect. 3.1) and two are by Ramanujan (Sects. 3.2, 3.3). Section 3.4 how to square the circle using the quadratrix.

The following table shows the formulas for the lengths that are constructed, their approximate values, the difference between these values and the value of $\pi$ and the error in meters that results if the approximation is used to compute the circumference of the earth given that its radius is 6378 km.

| Construction | Formula | Value | Difference | Error (m) |
|---|---|---|---|---|
| $\pi$ | – | 3.14159265359 | – | – |
| Kochański | $\sqrt{\dfrac{40}{3} - 2\sqrt{3}}$ | 3.14153333871 | $5.93 \times 10^{-5}$ | 757 |
| Ramanujan 1 | $\dfrac{355}{113}$ | 3.14159292035 | $2.67 \times 10^{-7}$ | 3.4 |
| Ramanujan 2 | $\left(9^2 + \dfrac{19^2}{22}\right)^{1/4}$ | 3.14159265258 | $1.01 \times 10^{-9}$ | 0.013 |

© The Author(s) 2022
M. Ben-Ari, *Mathematical Surprises*, https://doi.org/10.1007/978-3-031-13566-8_3

## 3.1 Kochański's Construction

**Construction (Fig. 3.1):**

1. Construct a unit circle centered at $O$, let $\overline{AB}$ be a diameter and construct a tangent to the circle at $A$.
2. Construct a unit circle centered at $A$ and denote its intersection with the first circle by $C$. Construct a unit circle centered at $C$ and denote its intersection with the second circle is $D$.
3. Construct $\overline{OD}$ and denote its intersection with the tangent by $E$.
4. From $E$ construct $F, G, H$, each at distance 1 from the previous point.
5. Construct $\overline{BH}$.

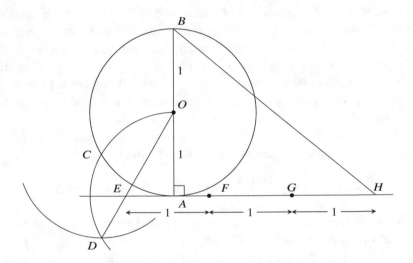

**Fig. 3.1** Kochański's approximation to $\pi$

**Theorem 3.1**    $\overline{BH} = \sqrt{\dfrac{40}{3} - 2\sqrt{3}} \approx \pi.$

**Proof** Figure 3.2 is an enlarged extract from Fig. 3.1, where dashed line segments have been added. Since all the circles are unit circles the lengths of the dashed lines are 1. It follows that $\overline{AOCD}$ is a rhombus so its diagonals are perpendicular to and bisect each other at the point labeled $K$. $\overline{AK} = 1/2$.

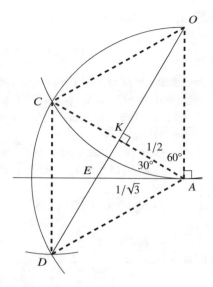

**Fig. 3.2** Detail from Kochański's construction

The diagonal $\overline{AC}$ forms two equilateral triangles $\triangle OAC, \triangle DAC$ so $\angle OAC = 60°$.
Since the tangent forms a right angle with the radius $\overline{OA}$, $\angle KAE = 30°$. Now:

$$\frac{1/2}{\overline{EA}} = \cos 30° = \frac{\sqrt{3}}{2}$$

$$\overline{EA} = \frac{1}{\sqrt{3}}$$

$$\overline{AH} = 3 - \overline{EA} = \left(3 - \frac{1}{\sqrt{3}}\right) = \frac{3\sqrt{3} - 1}{\sqrt{3}}.$$

$\triangle ABH$ is a right triangle and $\overline{AH} = 3 - \overline{EA}$, so by Pythagoras's Theorem:

$$\overline{BH}^2 = \overline{AB}^2 + \overline{AH}^2$$

$$= 4 + \frac{27 - 6\sqrt{3} + 1}{3} = \frac{40}{3} - 2\sqrt{3}$$

$$\overline{BH} = \sqrt{\frac{40}{3} - 2\sqrt{3}} \approx 3.141533387 \approx \pi.$$

$\square$

## 3.2 Ramanujan's First Construction

### Construction (Fig. 3.3):

1. Construct a unit circle centered at $O$ and let $\overline{PR}$ be a diameter.
2. Construct the point $H$ that bisects $\overline{PO}$ and the point $T$ that trisects $\overline{RO}$ (Thm. 2.6).
3. Construct the perpendicular at $T$ that intersects the circle at $Q$.
4. Construct the chords $\overline{RS} = \overline{QT}$ and $\overline{PS}$.
5. Construct a line parallel to $\overline{RS}$ from $T$ that intersects $\overline{PS}$ at $N$.
6. Construct a line parallel to $\overline{RS}$ from $O$ that intersects $\overline{PS}$ at $M$.
7. Construct the chord $\overline{PK} = \overline{PM}$.
8. Construct the tangent at $P$ of length $\overline{PL} = \overline{MN}$.
9. Connect the points $K, L, R$.
10. Find point $C$ such that $\overline{RC}$ is equal to $\overline{RH}$.
11. Construct $\overline{CD}$ parallel to $\overline{KL}$ that intersects $\overline{LR}$ at $D$.

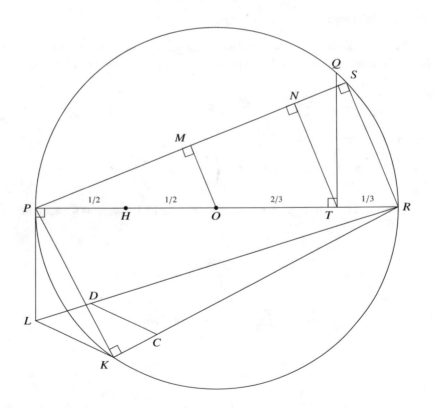

**Fig. 3.3** Ramanujan's first construction

**Theorem 3.2**  $\overline{RD}^2 = \dfrac{355}{113} \approx \pi.$

***Proof***  $\overline{RS} = \overline{QT}$ by construction and by Pythagoras's Theorem for $\triangle QOT$:

$$\overline{RS} = \overline{QT} = \sqrt{1^2 - \left(\frac{2}{3}\right)^2} = \frac{\sqrt{5}}{3}.$$

$\angle PSR$ is subtended by a diameter so $\triangle PSR$ is a right triangle. By Pythagoras's theorem:

$$\overline{PS} = \sqrt{2^2 - \left(\frac{\sqrt{5}}{3}\right)^2} = \sqrt{4 - \frac{5}{9}} = \frac{\sqrt{31}}{3}.$$

By construction $\overline{MO} \parallel \overline{RS}$ so $\triangle MPO \sim \triangle SPR$ and:

$$\frac{\overline{PM}}{\overline{PO}} = \frac{\overline{PS}}{\overline{PR}}$$

$$\frac{\overline{PM}}{1} = \frac{\sqrt{31}/3}{2}$$

$$\overline{PM} = \frac{\sqrt{31}}{6}.$$

By construction $\overline{NT} \parallel \overline{RS}$ so $\triangle NPT \sim \triangle SPR$ and:

$$\frac{\overline{PN}}{\overline{PT}} = \frac{\overline{PS}}{\overline{PR}}$$

$$\frac{\overline{PN}}{5/3} = \frac{\sqrt{31}/3}{2}$$

$$\overline{PN} = \frac{5\sqrt{31}}{18}$$

$$\overline{MN} = \overline{PN} - \overline{PM} = \sqrt{31}\left(\frac{5}{18} - \frac{1}{6}\right) = \frac{\sqrt{31}}{9}.$$

$\triangle PKR$ is a right triangle because $\angle PKR$ is subtended by a diameter. By construction $\overline{PK} = \overline{PM}$ and by Pythagoras's Theorem:

$$\overline{RK} = \sqrt{2^2 - \left(\frac{\sqrt{31}}{6}\right)^2} = \frac{\sqrt{113}}{6}.$$

$\triangle LPR$ is a right triangle because $\overline{PL}$ is a tangent so $\angle LPR$ is a right angle. $\overline{PL} = \overline{MN}$ by construction and by Pythagoras's Theorem:

$$\overline{RL} = \sqrt{2^2 + \left(\frac{\sqrt{31}}{9}\right)^2} = \frac{\sqrt{355}}{9}.$$

By construction $\overline{RC} = \overline{RH} = 3/2$ and $\overline{CD} \parallel \overline{LK}$. By similar triangles:

$$\frac{\overline{RD}}{\overline{RC}} = \frac{\overline{RL}}{\overline{RK}}$$

$$\frac{\overline{RD}}{3/2} = \frac{\sqrt{355}/9}{\sqrt{113}/6}$$

$$\overline{RD} = \sqrt{\frac{355}{113}}$$

$$\overline{RD}^2 = \frac{355}{113} \approx 3.14159292035 \approx \pi.$$

In Fig. 3.4 the line segments are labeled with their lengths.                         □

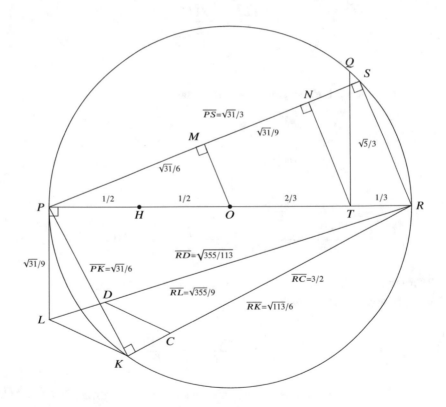

**Fig. 3.4** Ramanujan's first construction with labeled line segments

## 3.3 Ramanujan's Second Construction

**Construction (Fig. 3.5):**

1. Construct a unit circle centered at $O$ with diameter $\overline{AB}$ and let $C$ be the intersection of the perpendicular to $\overline{AB}$ at $O$ with the circle.
2. Trisect the line segment $\overline{AO}$ such that $\overline{AT} = 1/3$ and $\overline{TO} = 2/3$ (Thm. 2.6).
3. Construct $\overline{BC}$ and find points $M, N$ such that $\overline{CM} = \overline{MN} = \overline{AT} = 1/3$.
4. Construct $\overline{AM}, \overline{AN}$ and let $P$ be the point on $\overline{AN}$ such that $\overline{AP} = \overline{AM}$.
5. From $P$ construct a line parallel to $\overline{MN}$ that intersects $\overline{AM}$ at $Q$.
6. Construct $\overline{OQ}$ and then from $T$ construct a line parallel to $\overline{OQ}$ that intersects $\overline{AM}$ at $R$.
7. Construct $\overline{AS}$ tangent to $A$ such that $\overline{AS} = \overline{AR}$.
8. Construct $\overline{SO}$.

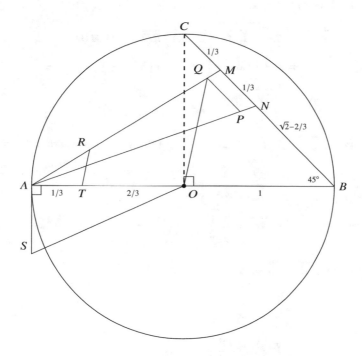

**Fig. 3.5** Ramanujan's second construction

**Theorem 3.3**   $3\sqrt{\overline{SO}} = \left(9^2 + \dfrac{19^2}{22}\right)^{1/4} \approx \pi.$

**Proof** $\triangle COB$ is a right triangle so by Pythagoras's Theorem $\overline{CB} = \sqrt{2}$ and:

$$\overline{NB} = \sqrt{2} - 2/3.$$

$\triangle COB$ is isoceles so $\angle NBA = \angle MBA = 45°$. By the Law of Cosines:

$$\overline{AN}^2 = \overline{BA}^2 + \overline{BN}^2 - 2 \cdot \overline{BA} \cdot \overline{BN} \cdot \cos \angle NBA$$

$$= 2^2 + \left(\sqrt{2} - \frac{2}{3}\right)^2 - 2 \cdot 2 \cdot \left(\sqrt{2} - \frac{2}{3}\right) \cdot \frac{\sqrt{2}}{2} = \frac{22}{9}$$

$$\overline{AN} = \sqrt{\frac{22}{9}}.$$

Again by the Law of Cosines:

$$\overline{AM}^2 = \overline{BA}^2 + \overline{BM}^2 - 2 \cdot \overline{BA} \cdot \overline{BM} \cdot \cos \angle MBA$$

$$= 2^2 + \left(\sqrt{2} - \frac{1}{3}\right)^2 - 2 \cdot 2 \cdot \left(\sqrt{2} - \frac{1}{3}\right) \cdot \frac{\sqrt{2}}{2} = \frac{19}{9}$$

$$\overline{AM} = \sqrt{\frac{19}{9}}.$$

By construction $\overline{QP} \parallel \overline{MN}$ so $\triangle MAN \sim \triangle QAP$, and by construction $\overline{AP} = \overline{AM}$:

$$\frac{\overline{AQ}}{\overline{AM}} = \frac{\overline{AP}}{\overline{AN}} = \frac{\overline{AM}}{\overline{AN}}$$

$$\overline{AQ} = \frac{\overline{AM}^2}{\overline{AN}} = \frac{19/9}{\sqrt{22/9}} = \frac{19}{3\sqrt{22}}.$$

By construction $\overline{TR} \parallel \overline{OQ}$ so $\triangle RAT \sim \triangle QAO$ and:

$$\frac{\overline{AR}}{\overline{AQ}} = \frac{\overline{AT}}{\overline{AO}}$$

$$\overline{AR} = \overline{AQ} \cdot \frac{\overline{AT}}{\overline{AO}} = \frac{19}{3\sqrt{22}} \cdot \frac{1/3}{1} = \frac{19}{9\sqrt{22}}.$$

By construction $\overline{AS} = \overline{AR}$ and $\triangle OAS$ is a right triangle because $\overline{AS}$ is a tangent. By Pythagoras's Theorem:

$$\overline{SO} = \sqrt{1^2 + \left(\frac{19}{9\sqrt{22}}\right)^2}$$

$$3\sqrt{\overline{SO}} = 3\left(1^2 + \frac{19^2}{9^2 \cdot 22}\right)^{1/4} = \left(9^2 + \frac{19^2}{22}\right)^{1/4} \approx 3.14159265258 \approx \pi.$$

In Fig. 3.6 the line segments are labeled with their lengths.  □

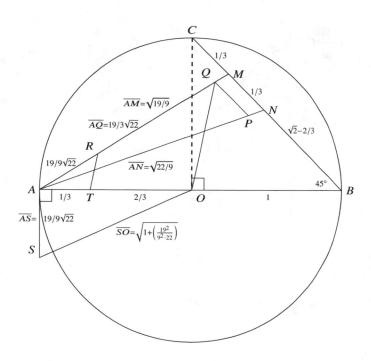

**Fig. 3.6** Ramanujan's second construction with labeled line segments

## 3.4 Squaring a Circle Using a Quadratrix

The quadratrix is described in Sect. 2.4.

Let $t = \overline{DE}$ be the distance that the horizontal straightedge has moved down the $y$-axis, and let $\theta$ be the corresponding angle between the rotating straightedge and the $x$-axis. Let $P$ the position of the joint of the two straightedges. The locus of $P$ is the quadratrix curve.

Let $F$ be the projection of $P$ onto the $x$-axis and let $G$ be the position of the joint when both straightedges reach the $x$-axis, that is, $G$ is the intersection of the quadratrix curve and the $x$-axis (Fig. 3.7).

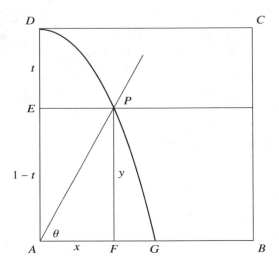

**Fig. 3.7** Squaring the circle with a quadratrix

**Theorem 3.4**    $\overline{AG} = 2/\pi$.

***Proof*** Let $y = \overline{PF} = \overline{EA} = 1 - t$. Since on a quadratrix $\theta$ decreases at the same rate that $t$ increases:

$$\frac{1-t}{1} = \frac{\theta}{\pi/2}$$

$$\theta = \frac{\pi}{2}(1 - t).$$

Let $x = \overline{AF} = \overline{EP}$. Then $\tan \theta = y/x$ so:

$$x = \frac{y}{\tan \theta} = y \cot \theta = y \cot \frac{\pi}{2}(1 - t) = y \cot \frac{\pi}{2} y. \tag{3.1}$$

We usually express a function as $y = f(x)$ but it can also be expressed as $x = f(y)$.

To obtain $x = \overline{AG}$ we can't simply plug $y = 0$ into Eq. 3.1, because $\cot 0$ is not defined, so let us compute the limit of $x$ as $y$ goes to 0. First perform the substitution $z = (\pi/2)y$ to obtain:

$$x = y \cot \frac{\pi}{2} y = \frac{2}{\pi}\left(\frac{\pi}{2} y \cot \frac{\pi}{2} y\right) = \frac{2}{\pi}(z \cot z),$$

and then take the limit:

$$\lim_{z \to 0} x = \frac{2}{\pi} \lim_{z \to 0}(z \cot z) = \frac{2}{\pi} \lim_{z \to 0}\left(\frac{z \cos z}{\sin z}\right) = \frac{2}{\pi} \lim_{z \to 0}\left(\frac{\cos z}{(\sin z)/z}\right) = \frac{2}{\pi} \frac{\cos 0}{1} = \frac{2}{\pi},$$

where we have used $\lim_{z \to 0}(\sin z/z) = 1$ (Thm. A.12). □

## What Is the Surprise?

It is surprising that such accurate approximations to $\pi$ can be constructed. Of course one can't help but be astonished by Ramanujan's clever constructions.

## Sources

Kochański's construction appears in [7]. Ramanujan's constructions are from [38, 39]. Squaring the circle using the quadratrix is from [31, pp. 48–49] and [62].

# Chapter 4
# The Five-Color Theorem

Maps use colors to distinguish one region from another by ensuring that adjacent regions are colored with different colors. In 1852 Francis Guthrie noticed that a map of the counties of England could be colored using only four countries. The claim that four countries suffice to color any planar map is called the *four-color theorem* and was only proved in 1976 by Kenneth Appel and Wolfgang Haken. They used sophisticated mathematical arguments to show that if there is a counterexample (a map needing more than four colors), it had to be associated with one of 1834 configurations. They then used a computer to check these configurations.

While the four-color theorem is extremely difficult to prove, the proofs of the five- and six-color theorems are relatively simple (Sects. 4.5, 4.6). On the way to proving these theorems, we define planar maps and graphs (Sect. 4.1), prove Euler's formula (Sect. 4.2) and show that a planar graph must have vertex whose degree is less than or equal to five. In Sect. 4.3 Euler's formula is used to show that two graphs are not planar.

In 1879 Alfred B. Kempe published a proof of the four-color theorem, but in 1890 Percy J. Heawood showed that the proof is incorrect. In Sect. 4.7 we present Kempe's flawed proof and Heawood's demonstration that it is not correct.

## 4.1 Planar Maps and Graphs

**Definition 4.1** A *planar map* is a set of regions in the plane separated by boundaries. A *coloring* of a map is an assignment of a color to each region such that regions sharing a boundary are assigned different colors.

Figure 4.1a shows a five-coloring of a planar map with ten regions. Figure 4.1b shows a four-coloring of the same map.

© The Author(s) 2022
M. Ben-Ari, *Mathematical Surprises*, https://doi.org/10.1007/978-3-031-13566-8_4

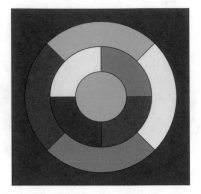

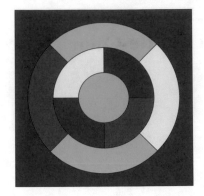

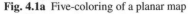

**Fig. 4.1a** Five-coloring of a planar map          **Fig. 4.1b** Four-coloring of a planar map

**Definition 4.2** A *graph* is a set of *vertices V* and a set of *edges E*, such that each edge is incident with exactly two vertices.

A *planar graph* is a graph such that no edges cross each other. In a planar graph, areas enclosed by a set of edges are called *faces*.

A *coloring* of a planar graph is an assignment of colors to vertices such that no two vertices of the same color are connected by an edge.

Planar maps and planar graphs are dual and it is convenient to investigate coloring problems in graphs rather than maps.

**Theorem 4.1** *Given a planar map, a planar graph can be constructed such that for each coloring of the regions of the map there is a coloring of the vertices of the graph, and conversely.*

**Proof** Construct one vertex for each region and construct an edge between two vertices if and only if the corresponding regions share a boundary.          □

*Example 4.1* Figure 4.2a shows the planar map from Fig. 4.1b and the vertices associated with the regions. Figure 4.2b shows the planar graph that corresponds to the map.

We can further limit our graphs to those whose faces are triangular.

**Definition 4.3** A graph is *triangular* if all its faces are bounded by three edges. A graph can be *triangulated* if edges can be added so that the graph is triangular. We also say that there is a *triangulation* of the graph.

*Example 4.2* The faces in the planar graph in Fig. 4.2b are triangular since each one is bounded by three edges. The edges are curved so the faces are not triangles, which are polygons whose three edges are straight line segments.

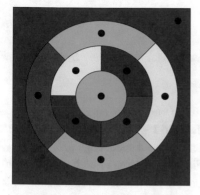

**Fig. 4.2a** Associating vertices with the regions of a planar map

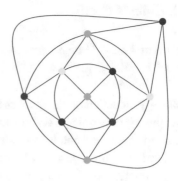

**Fig. 4.2b** The planar graph that corresponds to the planar map

**Fáry's Theorem** states that any triangular planar graph can be be transformed into an equivalent planar graph whose edges are straight line segments. Therefore, with no loss of generality, proofs can be restricted to planar graphs whose faces are triangles.

*Example 4.3* Fig. 4.3 (left) shows that a square can be two-colored, but if it is triangulated (center), four colors are necessary. Our goal is to prove that *all* graphs can be $n$-colored for some $n$. If the triangulated graph is $n$-colored, so is the original graph, because deleting the extra edges does not invalidate the coloring (right).

**Fig. 4.3** Coloring a triangulated graph

## 4.2  Euler's Formula

**Theorem 4.2** *Let G be a connected planar graph with V vertices, E edges and F faces. Then V − E + F = 2.*

**Proof** By induction on the number of edges. If the number of edges in the graph is zero, there is only a single vertex and a single face, so $1 - 0 + 1 = 2$. Otherwise, there is at least one edge $e$ and it connects two vertices $v_1, v_2$. Delete edge $e$.

*Case 1:* The graph becomes disconnected (Fig. 4.4a). Merge $v_1$ with $v_2$ (Fig. 4.4b). The resulting graph $G'$ is a planar connected graph and has fewer edges than $G$, so by the induction hypothesis $(V - 1) - (E - 1) + F = 2$ since the number of vertices is also reduced by one. Simplifying, we get $V - E + F = 2$ for $G$.

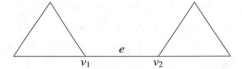

**Fig. 4.4a** Removing an edge disconnects the graph

**Fig. 4.4b** Merging two vertices

*Case 2:* The graph remains connected (Fig. 4.5a). $G'$ has fewer edges than $G$ (Fig. 4.5b), so by the induction hypothesis $V - (E - 1) + (F - 1) = 2$ since removing the edge joins two faces into one. Simplifying, we get $V - E + F = 2$ for $G$.      □

**Fig. 4.5a** Removing an edge does not disconnect the graph

**Fig. 4.5b** The graph remains connected and has fewer edges

**Theorem 4.3** *Let G be a connected, triangulated planar graph with E edges and V vertices. Then E = 3V − 6.*

**Proof** Each face is bounded by three edges, so $E = 3F/2$, where we divided by 2 because each edge has been counted twice, once for each face it bounds. By Euler's formula:

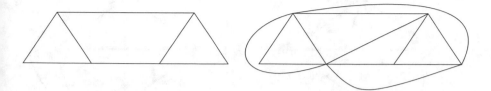

**Fig. 4.6a** Fewer edges than the upper limit

**Fig. 4.6b** In a triangulated graph the number of edges is maximal

$$E = V + F - 2$$
$$= V + 2E/3 - 2$$
$$= 3V - 6.$$

*Example 4.4* The planar graph in Fig. 4.2b has 10 vertices and $3 \cdot 10 - 6 = 24$ edges.

**Theorem 4.4** *Let G be a connected planar graph. Then $E \leq 3V - 6$.*

*Proof* Triangulate $G$ to obtain $G'$. $E' = 3V' - 6$ by Thm. 4.4. Now remove edges from $G'$ to obtain $G$. The number of vertices does not change so $E \leq 3V - 6$. □

*Example 4.5* The graph in Fig. 4.6a has 8 edges and 6 vertices and $8 < 3 \cdot 6 - 6 = 12$. Figure 4.6b shows a triangulated graph with 6 vertices and $3 \cdot 6 - 6 = 12$ edges.

## 4.3 Non-planar Graphs

Let us take a short detour to show how Thms. 4.2 and 4.4 can be used to prove that certain graphs are not planar.

**Theorem 4.5** $K_5$, *the complete graph on five vertices, is not planar (Fig. 4.7a).*

*Proof* For $K_5$, $V = 5$ and $E = 10$. By Thm. 4.4 the number of edges must be less than or equal to $3 \cdot 5 - 6 = 9$ so the graph is not planar. □

**Theorem 4.6** $K_{3,3}$, *the bipartite graph with three vertices on each side, is not planar (Fig. 4.8a).*

*Proof* $V = 6$ and $E = 9$. By Thm 4.2 if $K_{3,3}$ is planar, $F = E - V + 2 = 9 - 6 + 2 = 5$. But each face is bounded by four edges (Fig. 4.8b), so $E = 4F/2 = 10 \neq 9$. □

In 1930 Kazimierz Kuratowski proved a converse to these theorems: if a graph is not planar it contains (in a certain sense) $K_5$ or $K_{3,3}$.

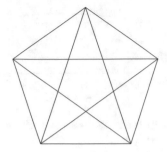

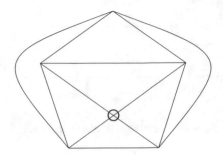

**Fig. 4.7a** $K_5$ is not planar

**Fig. 4.7b** A failed attempt to draw $K_5$ as planar

## 4.4 The Degrees of the Vertices

**Definition 4.4** $d(v)$, the *degree* of vertex $v$, is the number of edges incident with $v$.

*Example 4.6* The graph in Fig. 4.2b contains 8 vertices corresponding to the two rings and each vertex is of degree 5. The vertex corresponding to the outer face is of degree 4 as is the vertex corresponding to the inner face. Therefore:

$$\sum_{v \in V} d(v) = 5 \cdot 8 + 4 \cdot 2 = 48 \,.$$

To get the total number of edges divide 48 by 2 because each edge was counted twice, once for each of the vertices it is incident to.

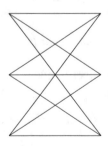

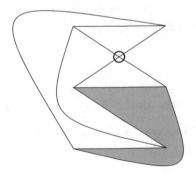

**Fig. 4.8a** $K_{3,3}$ is not planar

**Fig. 4.8b** A failed attempt to draw $K_{3,3}$ as planar

By generalizing the argument we get:

**Theorem 4.7** *Let $d_i$ for $i$ in $\{1, 2, 3, \ldots, k\}$ be the number of vertices of degree $i$ in a connected planar graph $G$ with $V$ vertices and $E$ edges, where $k$ is the highest degree of a vertex in $V$. Then:*

$$\sum_{v \in V} d(v) = \sum_{i=1}^{k} i \cdot d_i = 2E .$$

**Theorem 4.8** *Let $G$ be a connected planar graph with $E$ edges and $V$ vertices, and let $d_i$ for $i$ in $\{1, 2, 3, \ldots, k\}$ be the number of vertices of degree $i$, where $k$ is the highest degree of a vertex in $V$. Then there must be a vertex $v$ in $V$ such that $d(v) \leq 5$.*

**Proof (1)** If there are $d_1$ vertices of degree 1, $d_2$ vertices of degree 2, $\ldots$, $d_k$ vertices of degree $k$, then $V = \sum_{i=1}^{k} d_i$. From Thms. 4.4 and 4.7:

$$\sum_{i=1}^{k} i \cdot d_i = 2E \leq 2(3V - 6) = 6V - 12 = 6 \sum_{i=1}^{k} d_i - 12 .$$

Therefore:

$$\sum_{i=1}^{k} i \cdot d_i \leq 6 \sum_{i=1}^{k} d_i - 12$$

$$\sum_{i=1}^{k} (6 - i) d_i \geq 12 .$$

Since $12 > 0$ and all $d_i$ are positive, for least one $i$, $6 - i > 0$ and for that $i$, $i < 6$. □

**Proof (2)** Let us compute the *average* degree of the vertices which is the sum of the degrees divided by the number of vertices:

$$d_{avg} = \frac{\sum_{i=1}^{k} i \cdot d_i}{V} .$$

But the sum of the degrees is twice the number of edges which by Thm. 4.4 gives:

$$d_{avg} = \frac{2E}{V} \leq \frac{6V - 12}{V} = 6 - \frac{6}{V} < 6 .$$

If the average is less than six there must be a vertex of degree less than six.    □

*Example 4.7* In Fig. 4.2b the sum of the degrees is $8 \cdot 5 + 2 \cdot 4 = 48$. There are 10 vertices so the average degree is $48/10 = 4.8$ and there must be a vertex of degree 4 or less.

## 4.5  The Six-Color Theorem

**Theorem 4.9** *Any planar graph G can be six-colored.*

**Proof** By induction on the number of vertices. If $G$ has six vertices or fewer, six colors suffice. For the inductive step, by Thm. 4.8 $G$ has a vertex $v$ with degree 5 or fewer. Delete vertex $v$ to obtain the graph $G'$. By the induction hypothesis $G'$ can be six-colored, but $v$ has at most 5 neighbors and at most 5 colors are used to color them (Fig. 4.9a), so $v$ can be colored using the sixth color (Fig. 4.9b).        □

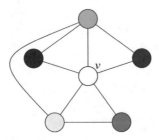      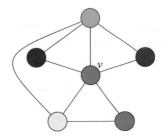

**Fig. 4.9a** Five colors suffice for coloring the neighbors of $v$                **Fig. 4.9b** Color $v$ with the sixth color

## 4.6  The Five-Color Theorem

**Definition 4.5** Let $G$ be a colored planar graph. A *(Kempe) chain* $G'$ is a maximal, two-colored, connected subgraph of $G$.

**Theorem 4.10** *Any planar graph G can be five-colored.*

**Proof** By induction on the number of vertices. If $G$ five vertices or fewer, five colors suffice. For the inductive step, by Thm. 4.8 $G$ has a vertex $v$ with degree 5 or less. Delete $v$ to obtain $G'$. By the induction hypothesis, $G'$ can be five-colored. In $G$, if the degree of $v$ is less than 5, or if $v_1, \ldots, v_5$, the neighbors of $v$, are colored with four colors or fewer, $v$ can be colored with the fifth color. Otherwise, $v_1, \ldots, v_5$ are colored with different colors in $G'$ (Fig. 4.10, top).

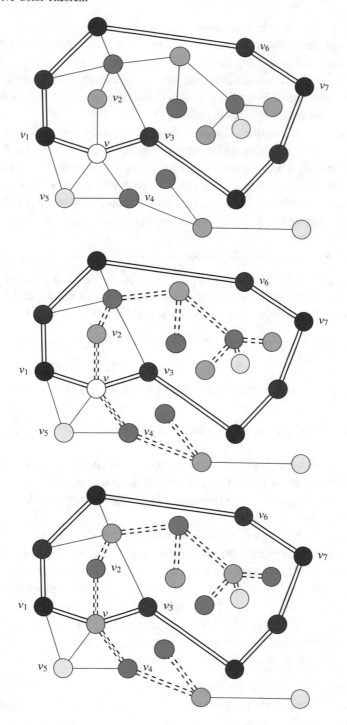

**Fig. 4.10** Proof of the five-color theorem

Consider vertex $v_1$ which is colored blue and vertex $v_3$ which is colored red. If $v_1, v_3$ are not connected by a blue-red path (say if the edge $\overline{v_6 v_7}$ did not exist), we can exchange the colors along the path from $v_1$ to $v_6$ and color $v$ blue. Otherwise, consider the blue-red chain which contains $v_1, v_3$. By adding $v$ and the edges $\overline{v v_1}, \overline{v v_3}$ we obtain a closed path $P$ (double line) that divides the plane into an "inside" region and an "outside" region (Fig. 4.10, middle)

Consider $v_2$ which is colored green and $v_4$ which is colored orange. These vertices *cannot* be contained in a single green-orange chain, because $v_2$ is *inside P* and $v_4$ is *outside P*, so any path connecting them must cross $P$, contradicting the assumption that the graph is planar. Therefore, they must be contained in two *unconnected* green-orange chains (double dashed line, in Fig. 4.10, middle). Exchange the colors on the chain containing $v_2$ and then $v$ can be colored green to obtain a five-coloring of $G$ (Fig. 4.10, bottom).                                                                      □

> The statement that a continuous path from the *inside* of of a closed continuous curve $P$ to the *outside* of $P$ must intersect $P$ is the **Jordan Curve Theorem**. The theorem is intuitively obvious but difficult to prove.

## 4.7 Kempe's Incorrect Proof of the Four-Color Theorem

**Theorem 4.11** *Any planar graph $G$ can be four-colored.*

***Proof (Incorrect)*** The base case of the induction and most of the proof is the same as that of the five-color theorem. The new case that must be considered is a vertex $v$ with five neighbors which, by the inductive hypothesis, can be colored with four colors after removing $v$.

In Fig. 4.11a there are two vertices $v_2, v_5$ colored blue. Consider the blue-green chain containing $v_2$ and the blue-yellow chain containing $v_5$. The blue-green chain is contained within the closed path defined by the red-yellow chain containing $v_1, v_3$ (double line) and the blue-yellow chain in contained within the closed path defined by the red-green chain containing $v_1, v_4$ (double dashed line).

Exchange the colors of both the blue-green chain and the blue-yellow chain (Fig. 4.11b). The result is that the neighbors of $v$ are colored with the three colors red, green and yellow, leaving blue free to color $v$.                                                                      □

Heawood noted that the closed paths defined by the red-yellow chain and the red-green chain can share red vertices ($v_1, v_8$ in Fig. 4.12a). When the colors are exchanged in the blue-green and blue-yellow chains, it is possible for blue vertices $v_6, v_7$ to be connected (Fig. 4.12b) and the coloring is no longer correct.

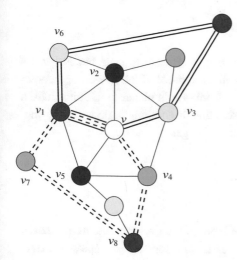

**Fig. 4.11a** Blue-green and blue-yellow Kempe chains

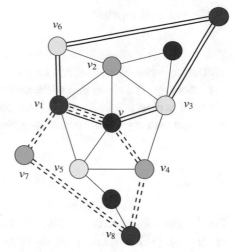

**Fig. 4.11b** Exchange the colors of the two Kempe chains

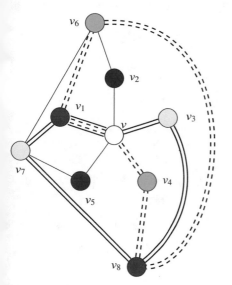

**Fig. 4.12a** Red-yellow and red-green chains share red vertices

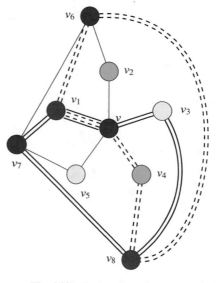

**Fig. 4.12b** Exchanging colors causes the blue vertices to become connected

## What Is the Surprise?

The four-color theorem is notorious because it is so easy to state but extremely difficult to prove. Therefore, it is surprising that the proof of the five-color theorem is elementary. The clever part of the proof is Thm. 4.8 (a planar graph must have a vertex of at most degree 5), which is a theorem that has nothing to do with coloring. Instead, it results just from counting vertices and edges.

## Sources

For the four-color theorem see [49, 54]. The proof of the five-color theorem is based on [1, 53]. [16] presents numerous proofs of Euler's formula. Kempe's incorrect proof of the four-color theorem is described in [46].

# Chapter 5
# How to Guard a Museum

In 1973 Victor Klee asked how many guards are need to observe all the walls of a museum? If the walls form a regular polygon or even a convex polygon, one guard is sufficient (Fig. 5.1).

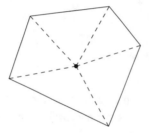

**Fig. 5.1** A museum whose walls form a convex polygon

Consider now a museum with saw-toothed walls (Fig. 5.2). Verify by counting that the museum has 15 walls. Each "tooth" defines a triangle that is shaded gray in Fig. 5.3. A guard placed anywhere within one of the triangles can observe all the walls bounding that triangle (red arrows).

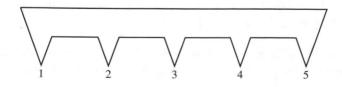

**Fig. 5.2** A museum whose walls do not form a convex polygon

© The Author(s) 2022
M. Ben-Ari, *Mathematical Surprises*, https://doi.org/10.1007/978-3-031-13566-8_5

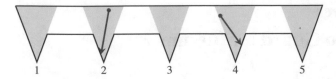

**Fig. 5.3** Visibility within each "tooth"

If at least one of the guards is placed near the top wall spanning the entire museum, she can observe all the horizontal walls (blue arrows in Fig. 5.4). Thus $5 = 15/3$ guards are sufficient to observe all the walls of the museum. Since the triangles do not overlap a guard in one triangle will not be able to observe all the walls of another triangle (green arrow) so 5 guards are necessary.

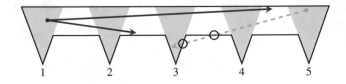

**Fig. 5.4** Visibility of the walls of the museum

The example in Fig. 5.2 can be generalized to $n/3$ teeth with $n$ walls, so we conclude that *at least* $n/3$ guards are necessary. We wish to prove that $n/3$ guards are sufficient to guard any museum.

Section 5.1 proves that any triangulated polygon can be three-colored. This is used in Sect. 5.2 to prove the theorem that $n/3$ guards are sufficient. Section 5.3 completes the proof by showing that any polygon can be triangulated.

## 5.1 Coloring Triangulated Polygons

**Definition 5.1** A *diagonal* a of polygon is an edge connecting two vertices that is not one of the (outside) edges of the polygon.

**Definition 5.2** A polygon can be *triangulated* if non-intersecting diagonals can be constructed such that the interior of the polygon is covered by triangles.

**Theorem 5.1** *Any polygon can be triangulated.*

We defer the proof of Thm. 5.1.

**Definition 5.3** A vertex of a polygon is *convex* if its interior angle is less than 180°; a vertex is *concave* if its interior angle is greater than 180°.

In Fig. 5.5 vertex 1 is convex and vertex 2 is concave.

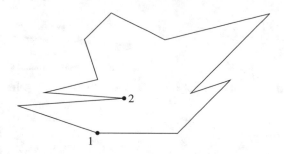

**Fig. 5.5** A polygon with a convex vertex (1) and a concave vertex (2)

**Definition 5.4** A polygon with vertices $V$ can be *three-colored* if there is a map:

$$c : V \mapsto \{red, blue, green\},$$

such that no edge has two vertices that are assigned the same color.

**Theorem 5.2** *A triangulated polygon can be three-colored.*

*Proof* By induction on the number of vertices. A triangle can be three-colored. A triangulated polygon with $n > 3$ vertices must have a diagonal. Choose an arbitrary diagonal $\overline{AB}$ (Fig. 5.6a) and divide the polygon along this diagonal into two smaller polygons (Fig. 5.6b). By induction each of these smaller polygons can be three-colored (Fig. 5.7a).

Since the colors assigned are arbitrary, if different colors are assigned to $A$, $B$ in the two polygons, we can rename the colors in one of them so that the colors of $A$, $B$ are the same in both polygons. For example, in Fig. 5.7b exchange *red* and *green* in the lower polygon. Paste the two polygons together to recover the original polygon with $n$ vertices. It will be three-colored (Fig. 5.8).                               □

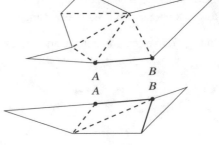

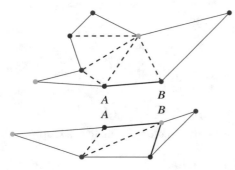

**Fig. 5.6a** An arbitrary diagonal in a polygon

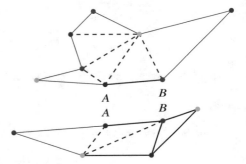

**Fig. 5.6b** Divide the polygon

**Fig. 5.7a** Three-color the two smaller polygons

**Fig. 5.7b** Exchange the colors of one polygon
to match the other

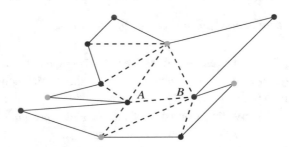

**Fig. 5.8** Paste the two smaller polygons back together

## 5.2 From Coloring of Polygons to Guarding a Museum

**Theorem 5.3** *A museum with n walls can be guarded by n/3 guards.*

***Proof*** By Thm. 5.1 the polygon can be triangulated and by Thm. 5.2 the polygon
can be three-colored. All three vertices of each triangle in the triangulation must be
colored by *different* colors in order to satisfy the condition of being three-colored.
Since the polygon is three-colored, at least one color, say red, can appear at most $n/3$

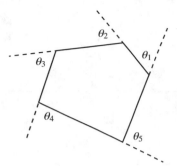

**Fig. 5.9** The exterior angles of a convex polygon

times, and every triangle must have a vertex colored red. Station a guard at each red vertex; she can observe all the walls of the each triangle the vertex belongs to. Since the triangles of the triangulation include all the edges of the polygon, $n/3$ guards are sufficient to observe all the walls of the museum. □

If $n$ is not divisible by 3 the number of guards needed is $\lfloor n/3 \rfloor$, the largest integer less than or equal to $n/3$. For example, 4 guards are sufficient for museums with 12, 13, 14 walls since $\lfloor 12/3 \rfloor = \lfloor 13/3 \rfloor = \lfloor 14/3 \rfloor = 4$. For simplicity we ignore this complication.

## 5.3 Any Polygon Can Be Triangulated

**Theorem 5.4** *The sum of the interior angles of a polygon with n vertices is:*

$$180°(n-2).$$

**Proof** Consider a convex polygon and denote its *exterior angles* by $\theta_i$ (Fig. 5.9). As you move from one dashed line in sequence to the next dashed line, you complete a rotation around a circle so:

$$\sum_{i=1}^{n} \theta_i = 360°.$$

For each exterior angle $\theta_i$ denote its corresponding interior angle by $\phi_i$. Then:

$$\sum_{i=1}^{n} \theta_i = \sum_{i=1}^{n}(180° - \phi_i) = 360°$$

$$\sum_{i=1}^{n} \phi_i = n \cdot 180° - 360° = 180°(n-2).$$

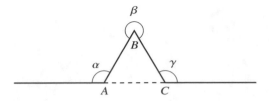

**Fig. 5.10** A concave vertex

If there is a concave vertex ($B$ in Fig. 5.10), there is a triangle formed by the two edges incident with the concave vertex and the line $\overline{AC}$ connecting the other two vertices. By summing the angles of the triangle we get:

$$(180° - \alpha) + (360° - \beta) + (180° - \gamma) = 180°$$
$$\alpha + \beta + \gamma = 3 \cdot 180° \,.$$

The sum of the interior angles increases by $\alpha + \beta + \gamma$ while the number of vertices increases by three preserving the equation in the theorem:

$$\sum_{i=1}^{n} \phi_i + (\alpha + \beta + \gamma) = 180°(n - 2) + 3 \cdot 180°$$
$$= 180°((n + 3) - 2) \,. \qquad \square$$

**Theorem 5.5** *There must be at least three convex vertices in a polygon.*

**Proof** Let $k$ be the number of concave vertices where the interior angle of each is $180° + \epsilon_i$, $\epsilon_i > 0$. The sum of the interior angles of the *concave* vertices is certainly less than or equal to the sum of the interior angles of *all* the vertices:

$$k \cdot 180° + \sum_{i=1}^{k} \epsilon_i \leq 180°(n - 2)$$
$$(k + 2) \cdot 180° + \sum_{i=1}^{k} \epsilon_i \leq n \cdot 180°$$
$$(k + 2) \cdot 180° < n \cdot 180°$$
$$k < n - 2 \,.$$

It follows that there must at least three vertices that are convex, not concave.  $\square$

**Proof (Theorem 5.1)** By induction on the number of vertices. For $n = 3$ there is nothing to prove. If $n > 3$, by Thm. 5.5 there must be a convex vertex $C$. Label its adjacent vertices by $B, D$. If $\overline{BD}$ is contained within the polygon (Fig. 5.11a), it is

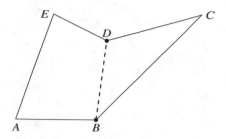

**Fig. 5.11a** Triangulation where a diagonal is contained within the polygon

**Fig. 5.11b** Triangulation where a diagonal is not contained within the polygon

a diagonal and the polygon can be split into $\triangle BCD$ and another polygon $\overline{ABDE}$ with $\overline{BD}$ as an edge and which is smaller than the original polygon (Fig. 5.11a). By the inductive hypothesis, the polygon can be triangulated and then pasted back to $\triangle BCD$, triangulating the original polygon.

If $\overline{BD}$ is not contained in the polygon, there must be concave vertex $F$ that is *closest* to $C$ (Fig. 5.11b). $\overline{CF}$ is a diagonal and splits the polygon into two smaller polygons $\overline{CFED}$ and $\overline{CFAB}$. By the inductive hypothesis these can be triangulated and pasted together. □

## What Is the Surprise?

The museum theorem is suprising because what seems to be a theorem in geometry is proved rather elegantly by an appeal to coloring a graph.

## Sources

This chapter is based on [1, Chap. 39].

# Chapter 6
# Induction

The axiom of mathematical induction is used extensively as a method of proof in mathematics. This chapter presents inductive proofs of results that may not be known to the reader. We begin with a short review of mathematical induction (Sect. 6.1). Section 6.2 proves results about the familiar Fibonacci numbers while Sect. 6.3 proves results about Fermat numbers. Section 6.4 presents the 91-function discovered by John McCarthy; the proof is by induction on an unusual sequence: integers in an inverse ordering. The proof of the formula for the Josephus problem (Sect. 6.5) is also unusual because of the double induction on two different parts of an expression.

## 6.1 The Axiom of Mathematical Induction

Mathematical induction is the primary method of proving statements to be true for an unbounded set of numbers. Consider:

$$1 = 1, \quad 1 + 2 = 3, \quad 1 + 2 + 3 = 6, \quad 1 + 2 + 3 + 4 = 10.$$

We might notice that:

$$1 = (1 \cdot 2)/2, \quad 3 = (2 \cdot 3)/2, \quad 6 = (3 \cdot 4)/2, \quad 10 = (4 \cdot 5)/2,$$

and then conjecture that for *all* integers $n \geq 1$:

$$\sum_{i=1}^{n} i = \frac{n(n+1)}{2}.$$

If you have enough patience, checking this formula for any specific value of $n$ is easy, but how can it be proved for *all* of the infinite number of positive integers? This is where mathematical induction comes in.

© The Author(s) 2022
M. Ben-Ari, *Mathematical Surprises*, https://doi.org/10.1007/978-3-031-13566-8_6

**Axiom 6.1** Let $P(n)$ be a property (such as an equation, a formula, or a theorem), where $n$ is a positive integer. Suppose that you can:

- *Base case*: Prove that $P(1)$ is true.
- *Inductive step*: For arbitrary $m$, prove that $P(m + 1)$ is true provided that you assume that $P(m)$ is true.

Then $P(n)$ is true for all $n \geq 1$. The assumption that $P(m)$ is true for arbitrary $m$ is called the *inductive hypothesis*.

Here is a simple example of a proof by mathematical induction.

**Theorem 6.2** *For $n \geq 1$:*

$$\sum_{i=1}^{n} i = \frac{n(n+1)}{2} .$$

**Proof** The base case is trivial:

$$\sum_{i=1}^{1} i = 1 = \frac{1(1+1)}{2} .$$

The inductive hypothesis is that the following equation is true for $m$:

$$\sum_{i=1}^{m} i = \frac{m(m+1)}{2} .$$

The inductive step is to prove the equation for $m + 1$:

$$\sum_{i=1}^{m+1} i = \sum_{i=1}^{m} i + (m + 1)$$
$$= \frac{m(m+1)}{2} + (m + 1) = \frac{(m+1)(m+2)}{2} .$$

By the principle of mathematical induction, for any $n \geq 1$:

$$\sum_{i=1}^{n} i = \frac{n(n+1)}{2} .$$

$\square$

The inductive hypothesis can be confusing because it seems that we are assuming what we are trying to prove. The reasoning is *not* circular because we assume the truth of a property for something *small* and then use the assumption to prove the property for something *larger*.

Mathematical induction is an axiom so there can be no question of proving induction. You just have to accept induction like you accept other axioms of mathematics such as $x + 0 = x$. Of course, you are free to reject mathematical induction, but then you will have to reject much of modern mathematics.

> Mathematical induction is a rule of inference that is one of the *Peano axioms*
> for formalizing natural numbers. The *well-ordering axiom* can be used to
> prove the axiom of induction and, conversely, the axiom of induction can
> be used to prove the well-ordering axiom. However, the axiom of induction
> cannot be proved from the other, more elementary, Peano axioms.

## 6.2 Fibonacci Numbers

Fibonacci numbers are the classic example of a recursive definition:

$$f_1 = 1$$
$$f_2 = 1$$
$$f_n = f_{n-1} + f_{n-2} \text{ for } n \geq 3 .$$

The first twelve Fibonacci numbers are:

$$1, 1, 2, 3, 5, 8, 13, 21, 34, 55, 89, 144 .$$

**Theorem 6.3** *Every fourth Fibonacci number is divisible by* 3.

*Example 6.1* $f_4 = 3 = 3 \cdot 1$, $f_8 = 21 = 3 \cdot 7$, $f_{12} = 144 = 3 \cdot 48$.

**Proof** Base case: $f_4 = 3$ is divisible by 3. The inductive hypothesis is that $f_{4n}$ is
divisible by 3. The inductive step is:

$$
\begin{aligned}
f_{4(n+1)} &= f_{4n+4} \\
&= f_{4n+3} + f_{4n+2} \\
&= (f_{4n+2} + f_{4n+1}) + f_{4n+2} \\
&= ((f_{4n+1} + f_{4n}) + f_{4n+1}) + f_{4n+2} \\
&= ((f_{4n+1} + f_{4n}) + f_{4n+1}) + (f_{4n+1} + f_{4n}) \\
&= 3 f_{4n+1} + 2 f_{4n} .
\end{aligned}
$$

$3 f_{4n+1}$ is divisible by 3 and, by the inductive hypothesis, $f_{4n}$ is divisible by 3.
Therefore, $f_{4(n+1)}$ is divisible by 3.                                                  □

**Theorem 6.4** $f_n < \left(\dfrac{7}{4}\right)^n$ .

***Proof*** Base cases: $f_1 = 1 < \left(\dfrac{7}{4}\right)^1$ and $f_2 = 1 < \left(\dfrac{7}{4}\right)^2 = \dfrac{49}{16}$. The inductive step is:

$$
\begin{aligned}
f_{n+1} &= f_n + f_{n-1} \\
&< \left(\frac{7}{4}\right)^n + \left(\frac{7}{4}\right)^{n-1} \\
&= \left(\frac{7}{4}\right)^{n-1} \cdot \left(\frac{7}{4} + 1\right) \\
&< \left(\frac{7}{4}\right)^{n-1} \cdot \left(\frac{7}{4}\right)^2 \\
&= \left(\frac{7}{4}\right)^{n+1} ,
\end{aligned}
$$

since:

$$
\left(\frac{7}{4} + 1\right) = \frac{11}{4} = \frac{44}{16} < \frac{49}{16} = \left(\frac{7}{4}\right)^2 .
$$

$\square$

**Theorem 6.5 (Binet's formula)**

$$
f_n = \frac{\phi^n - \bar{\phi}^n}{\sqrt{5}}, \quad \text{where } \phi = \frac{1 + \sqrt{5}}{2}, \ \bar{\phi} = \frac{1 - \sqrt{5}}{2} .
$$

***Proof*** We first show that $\phi^2 = \phi + 1$:

$$
\begin{aligned}
\phi^2 &= \left(\frac{1 + \sqrt{5}}{2}\right)^2 \\
&= \frac{1}{4} + \frac{2\sqrt{5}}{4} + \frac{5}{4} = \left(\frac{1}{2} + \frac{\sqrt{5}}{2}\right) + 1 \\
&= \phi + 1 .
\end{aligned}
$$

Similarly, we can show that $\bar{\phi}^2 = \bar{\phi} + 1$.
The base case for Binet's formula is:

$$
\frac{\phi^1 - \bar{\phi}^1}{\sqrt{5}} = \frac{\frac{1+\sqrt{5}}{2} - \frac{1-\sqrt{5}}{2}}{\sqrt{5}} = \frac{\sqrt{5}}{\sqrt{5}} = 1 = f_1 .
$$

Assume the inductive hypothesis for all $k \leq n$. The inductive step is:

$$\phi^{n+1} - \bar{\phi}^{n+1} = \phi^2 \phi^{n-1} - \bar{\phi}^2 \bar{\phi}^{n-1}$$
$$= (\phi + 1)\phi^{n-1} - (\bar{\phi} + 1)\bar{\phi}^{n-1}$$
$$= (\phi^n - \bar{\phi}^n) + (\phi^{n-1} - \bar{\phi}^{n-1})$$
$$= \sqrt{5}f_n + \sqrt{5}f_{n-1}$$

$$\frac{\phi^{n+1} - \bar{\phi}^{n+1}}{\sqrt{5}} = f_n + f_{n-1} = f_{n+1}. \qquad \square$$

**Theorem 6.6**

$$f_n = \binom{n}{0} + \binom{n-1}{1} + \binom{n-2}{2} + \cdots.$$

***Proof*** Let us first prove Pascal's rule:

$$\binom{n}{k} + \binom{n}{k+1} = \binom{n+1}{k+1}.$$

$$\binom{n}{k} + \binom{n}{k+1} = \frac{n!}{k!(n-k)!} + \frac{n!}{(k+1)!(n-(k+1))!}$$

$$= \frac{n!(k+1)}{(k+1)!(n-k)!} + \frac{n!(n-k)}{(k+1)!(n-k)!}$$

$$= \frac{n!(n+1)}{(k+1)!(n-k)!}$$

$$= \frac{(n+1)!}{(k+1)!((n+1)-(k+1))!}$$

$$= \binom{n+1}{k+1}.$$

We will also use the equality $\binom{k}{0} = \frac{k!}{0!(k-0)!} = 1$ for any $k \geq 1$.

We can now prove the theorem. The base case is:

$$f_1 = \binom{1}{0} = \frac{1!}{0!(1-0)!} = 1.$$

The inductive step is:

$$f_n = f_{n-1} + f_{n-2} = \binom{n-1}{0} + \binom{n-2}{1} + \binom{n-3}{2} + \binom{n-4}{3} + \cdots$$

$$\binom{n-2}{0} + \binom{n-3}{1} + \binom{n-4}{2} + \cdots$$

$$= \binom{n-1}{0} + \binom{n-1}{1} + \binom{n-2}{2} + \binom{n-3}{3} + \cdots$$

$$= \binom{n}{0} + \binom{n-1}{1} + \binom{n-2}{2} + \binom{n-3}{3} + \cdots . \qquad \square$$

## 6.3 Fermat Numbers

**Definition 6.1** The integers $F_n = 2^{2^n} + 1$ for $n \geq 0$ are called *Fermat numbers*.

The first five Fermat numbers are prime:

$$F_0 = 3, \quad F_1 = 5, \quad F_2 = 17, \quad F_3 = 257, \quad F_4 = 65537 .$$

The seventeenth-century mathematician Pierre de Fermat claimed that all Fermat numbers are prime, but nearly a hundred years later Leonhard Euler showed that:

$$F_5 = 2^{2^5} + 1 = 2^{32} + 1 = 4294967297 = 641 \times 6700417 .$$

Fermat numbers become extremely large as $n$ increases. It is known that Fermat numbers are not prime for $5 \leq n \leq 32$, but the factorization of some of those numbers is still not known.

**Theorem 6.7** *For $n \geq 2$, the last digit of $F_n$ is 7.*

***Proof*** The base case is $F_2 = 2^{2^2} + 1 = 17$. The inductive hypothesis is $F_n = 10k_n + 7$ for some $k_n \geq 1$. The inductive step is:

$$F_{n+1} = 2^{2^{n+1}} + 1 = 2^{2^n \cdot 2^1} + 1 = \left(2^{2^n}\right)^2 + 1$$

$$= \left(\left(2^{2^n} + 1\right) - 1\right)^2 + 1 = (F_n - 1)^2 + 1$$

$$= (10k_n + 7 - 1)^2 + 1 = (10k_n + 6)^2 + 1$$

$$= 100k_n^2 + 120k_n + 36 + 1$$

$$= 10(10k_n^2 + 12k_n + 3) + 6 + 1$$

$$= 10k_{n+1} + 7, \quad \text{for some } k_{n+1} \geq 1 . \qquad \square$$

**Theorem 6.8** *For $n \geq 1$, $F_n = \displaystyle\prod_{k=0}^{n-1} F_k + 2$.*

*Proof* The base case is:

$$F_1 = \prod_{k=0}^{0} F_k + 2 = F_0 + 2 = 3 + 2 = 5.$$

The inductive step is:

$$\prod_{k=0}^{n} F_k = \left( \prod_{k=0}^{n-1} F_k \right) F_n$$
$$= (F_n - 2) F_n$$
$$= \left( 2^{2^n} + 1 - 2 \right) \left( 2^{2^n} + 1 \right)$$
$$= \left( 2^{2^n} \right)^2 - 1 = \left( 2^{2^{n+1}} + 1 \right) - 2$$
$$= F_{n+1} - 2$$
$$F_{n+1} = \prod_{k=0}^{n} F_k + 2.$$

$\square$

## 6.4 McCarthy's 91-function

We usually associate induction with proofs of properties defined on the set of positive integers. Here we bring an inductive proof based on a strange ordering where larger numbers are less than smaller numbers. The induction works because the only property required of the set is that it be ordered under some relational operator.

Consider the following recursive function defined on the intergers:

$$f(x) = \text{if } x > 100 \text{ then } x - 10 \text{ else } f(f(x+11)).$$

For numbers greater than 100 the result of applying the function is trivial:

$$f(101) = 91, \quad f(102) = 92, \quad f(103) = 93, \quad f(104) = 94, \ \ldots .$$

What about numbers less than or equal to 100? Let us compute $f(x)$ for some numbers, where the computation in each line uses the results of previous lines:

$$
\begin{aligned}
f(100) &= f(f(100+11)) = f(f(111)) = f(101) = 91 \\
f(99) &= f(f(99+11)) = f(f(110)) = f(100) = 91 \\
f(98) &= f(f(98+11)) = f(f(109)) = f(99) = 91
\end{aligned}
$$

$$\cdots$$

$$
\begin{aligned}
f(91) &= f(f(91+11)) = f(f(102)) = f(92) \\
&= f(f(103)) = f(93) = \cdots = f(98) = 91 \\
f(90) &= f(f(90+11)) = f(f(101)) = f(91) = 91 \\
f(89) &= f(f(89+11)) = f(f(100)) = f(91) = 91 \,.
\end{aligned}
$$

Define the function $g$ as:

$$g(x) = \text{if } x > 100 \text{ then } x - 10 \text{ else } 91 \,.$$

**Theorem 6.9** *For all $x$, $f(x) = g(x)$.*

**Proof** The proof is by induction over the set of integers $S = \{x \mid x \le 101\}$ using the relational operator $\prec$ defined by:

$$y \prec x \text{ if and only if } x < y \,,$$

where on the right-hand side $<$ is the usual relational operator on the integers. This definition results in the following ordering:

$$101 \prec 100 \prec 99 \prec 98 \prec 97 \prec \cdots \,.$$

There are three cases to the proof. We use the results of the above computations.
 *Case 1:* $x > 100$. This is trivial by the definitions of $f$ and $g$.
 *Case 2:* $90 \le x \le 100$. The base case of the induction is:

$$f(100) = 91 = g(100) \,,$$

since we showed that $f(100) = 91$ and by definition $g(100) = 91$.
 The inductive assumption is $f(y) = g(y)$ for $y \prec x$ and the inductive step is:

$$
\begin{aligned}
f(x) &= f(f(x+11)) & \text{(6.1a)} \\
&= f(x + 11 - 10) = f(x+1) & \text{(6.1b)} \\
&= g(x+1) & \text{(6.1c)} \\
&= 91 & \text{(6.1d)} \\
&= g(x) \,. & \text{(6.1e)}
\end{aligned}
$$

Equation 6.1a holds by definition of $f$ since $x \le 100$. The equality of Eq. 6.1a and Eq. 6.1b holds by the definition of $f$, because $x \ge 90$ so $x + 11 > 100$. The equality of Eq. 6.1b and Eq. 6.1c follows by the inductive hypothesis $x \leq 100$, so $x + 1 \leq 101$ which implies that $x + 1 \in S$ and $x + 1 < x$. The equality of Eq. 6.1c, Eq. 6.1d and Eq. 6.1e follows by definition of $g$ and $x + 1 \leq 101$, so $x \leq 100$.

*Case 3:* $x < 90$. The base case is: $f(89) = f(f(100)) = f(91) = 91 = g(89)$ by definition of $g$ since $89 < 100$.

The inductive assumption is $f(y) = g(y)$ for $y < x$ and the inductive step is:

$$f(x) = f(f(x + 11)) \tag{6.2a}$$

$$= f(g(x + 11)) \tag{6.2b}$$

$$= f(91) \tag{6.2c}$$

$$= 91 \tag{6.2d}$$

$$= g(x). \tag{6.2e}$$

Equation 6.2a holds by definition of $f$ and $x < 90 \leq 100$. The equality of Eq. 6.2a and Eq 6.2b follows from the inductive hypothesis $x < 90$, so $x + 11 < 101$, which implies that $x + 11 \in S$ and $x + 11 < x$. The equality of Eq. 6.2b and Eq 6.2c follows by definition of $g$ and $x + 11 < 101$. Finally, we have already shown that $f(91) = 91$ and $g(x) = 91$ for $x < 90$ by definition.                                           □

## 6.5  The Josephus Problem

Josephus was the commander of the city of Yodfat during the Jewish rebellion against the Romans. The overwhelming strength of the Roman army eventually crushed the city's resistance and Josephus took refuge in a cave with some of his men. They preferred to commit suicide rather than being killed or captured by the Romans. According the account by Josephus, he arranged to save himself, became an observer with the Romans and later wrote a history of the rebellion. We present the problem as an abstract mathematical one.

**Definition 6.2 (Josephus problem)** Consider the numbers $1, \ldots, n+1$ arranged in a circle. Delete every $q$'th number going around the circle $q, 2q, 3q, \ldots$ (where the computation is performed modulo $n + 1$) until only one number $m$ remains. $J(n + 1, q) = m$ is the *Josephus number* for $n + 1$ and $q$.

*Example 6.2* Let $n + 1 = 41$ and let $q = 3$. Arrange the numbers in a circle:

$$\rightarrow \quad 1 \ 2 \ 3 \ 4 \ 5 \ 6 \ 7 \ 8 \ 9 \ 10 \ 11 \ 12 \ 13 \ 14 \ 15 \ 16 \ 17 \ 18 \ 19 \ 20 \ 21 \quad \downarrow$$
$$\uparrow \quad 41 \ 40 \ 39 \ 38 \ 37 \ 36 \ 35 \ 34 \ 33 \ 32 \ 31 \ 30 \ 29 \ 28 \ 27 \ 26 \ 25 \ 24 \ 23 \ 22 \quad \leftarrow$$

The first round of deletions leads to:

$\rightarrow$  1  2  $\cancel{3}$  4  5  $\cancel{6}$  7  8  $\cancel{9}$  10  11  $\cancel{12}$  13  14  $\cancel{15}$  16  17  $\cancel{18}$  19  20  $\cancel{21}$  $\downarrow$

$\uparrow$    41  40  $\cancel{39}$  38  37  $\cancel{36}$  35  34  $\cancel{33}$  32  31  $\cancel{30}$  29  28  $\cancel{27}$  26  25  $\cancel{24}$  23  22    $\leftarrow$

After removing the deleted numbers this can be written as:

1 2 4 5 7 8 10 11 13 14 16 17 19 20 22 23 25 26 28 29 31 32 34 35 37 38 40 41

The second round of deletions (starting at the last deletion of 39) leads to:

$\cancel{1}$ 2 4 $\cancel{5}$ 7 8 $\cancel{10}$ 11 13 $\cancel{14}$ 16 17 $\cancel{19}$ 20 22 $\cancel{23}$ 25 26 $\cancel{28}$ 29 31 $\cancel{32}$ 34 35 $\cancel{37}$ 38 40 $\cancel{41}$

We continue deleting every third number until only one remains:

2  4  $\cancel{7}$  8 11 $\cancel{13}$ 16 17 $\cancel{20}$ 22 25 $\cancel{26}$ 29 31 $\cancel{34}$ 35 38 $\cancel{40}$

2  4  $\cancel{8}$ 11 16 $\cancel{17}$ 22 25 $\cancel{29}$ 31 35 $\cancel{38}$

2  4 $\cancel{11}$ 16 22 $\cancel{25}$ 31 35

$\cancel{2}$  4 16 $\cancel{22}$ 31 35

$\cancel{4}$ 16 31 $\cancel{35}$

$\cancel{16}$ 31

31

It follows that $J(41, 3) = 31$.

The reader is invited to perform the computation for deleting every seventh number from a circle of 40 numbers in order to verify that the last number is 30.

**Theorem 6.10** $J(n + 1, q) = (J(n, q) + q) \pmod{n + 1}$.

*Proof* The first number deleted in the first round is the $q$'th number and the numbers that remain after the deletion are the $n$ numbers:

$$1 \; 2 \; \ldots \; q - 1 \; q + 1 \; \ldots \; n \; n + 1 \pmod{n + 1}.$$

Counting to find the next deletion starts with $q + 1$. Mapping $1, \ldots, n$ into this sequence we get:

| 1 | 2 | $\ldots n - q$ | $n + 1 - q$ | $n + 2 - q \ldots n - 1$ | $n$ | $\pmod{n+1}$ |
|---|---|---|---|---|---|---|
| $\downarrow$ | $\downarrow$ | $\downarrow$ | $\downarrow$ | $\downarrow$ | $\downarrow$ | $\downarrow$ |
| $q + 1$ | $q + 2 \ldots$ | $n$ | $n + 1$ | $1 \quad \ldots q - 2$ | $q - 1$ | $\pmod{n+1}$ . |

Remember that the computations are modulo $n + 1$:

$$\begin{aligned}(n + 2 - q) + q &= (n + 1) + 1 && = 1 && \pmod{n + 1} \\ (n) + q &&= (n + 1) - 1 + q &= q - 1 && \pmod{n + 1}.\end{aligned}$$

This is the Josephus problem for $n$ numbers, except that the numbers are offset by $q$. It follows that:

$$J(n+1, q) = (J(n, q) + q) \pmod{n+1}.$$ □

**Theorem 6.11** *For $n \geq 1$ there exist numbers $a \geq 0, 0 \leq t < 2^a$, such that $n = 2^a + t$.*

**Proof** This can be proved from repeated application of the division algorithm with divisors $2^0, 2^1, 2^2, 2^4, \ldots$, but it is easy to see from the binary representation of $n$. For some $a$ and $b_{a-1}, b_{a-2}, \ldots, b_1, b_0$, where for all $i$, $b_i = 0$ or $b_i = 1$, $n$ can be expressed as:

$$n = 2^a + b_{a-1}2^{a-1} + \cdots + b_0 2^0$$
$$n = 2^a + (b_{a-1}2^{a-1} + \cdots + b_0 2^0)$$
$$n = 2^a + t, \quad \text{where } t \leq 2^a - 1.$$ □

We now prove that there is simple closed form for $J(n, 2)$.

**Theorem 6.12** *For $n = 2^a + t$, $a \geq 0, 0 \leq t < 2^a$, $J(n, 2) = 2t + 1$.*

**Proof** By Thm. 6.11, $n$ can be expressed as stated in the theorem. The proof that $J(n, 2) = 2t + 1$ is by a double induction, first on $a$ and then on $t$.

*First induction:*

Base case. Assume that $t = 0$ so that $n = 2^a$. Let $a = 1$ so that there are two numbers in the circle $1, 2$. Since $q = 2$, the second number will be deleted, so the remaining number is 1 and $J(2^1, 2) = 1$.

The inductive hypothesis is that $J(2^a, 2) = 1$. What is $J(2^{a+1}, 2)$? In the first round all the even numbers are deleted:

$$1 \quad \cancel{2} \quad 3 \quad \cancel{4} \quad \ldots \quad 2^{a+1}-1 \quad \cancel{2^{a+1}}.$$

There are now $2^a$ numbers left:

$$1 \quad 3 \quad \ldots \quad 2^{a+1}-1.$$

By the inductive hypothesis $J(2^{a+1}, 2) = J(2^a, 2) = 1$ so by induction $J(n, 2) = 1$ whenever $n = 2^a + 0$.

*Second induction:*

We have proved $J(2^a + 0, 2) = 2 \cdot 0 + 1$, the base case of the second induction. The inductive hypothesis is $J(2^a + t, 2) = 2t + 1$. By Thm. 6.10:

$$J(2^a + (t + 1), 2) = J(2^a + t, 2) + 2 = 2t + 1 + 2 = 2(t + 1) + 1.$$ □

Theorems 6.11 and 6.12 give a simple algorithm for computing $J(n, 2)$. From the proof of Thm. 6.11:

$$n = 2^a + t = 2^a + (b_{a-1}2^{a-1} + \cdots + b_0 2^0),$$

so $t = b_{a-1}2^{a-1} + \cdots + b_0 2^0$. We simply multiply by 2 (shift left by one digit) and add 1. For example, since $n = 41 = 2^5 + 2^3 + 2^0 = 101001$, it follows that $J(41, 2) = 2t + 1$, and in binary notation:

$$41 = 101001$$
$$9 = 01001$$
$$2t + 1 = 10011 = 16 + 2 + 1 = 19.$$

The reader can verify the result by deleting every second number in a circle $1, \ldots, 41$.

There is a closed form for $J(n, 3)$ but it is quite complicated.

## What Is the Surprise?

Induction is perhaps the most important proof technique in modern mathematics. While Fibonacci numbers are extremely well-known and Fermat numbers are also easy to understand, I was surprised to find so many formulas that I never knew (such as Thms. 6.3 and 6.4) that can be proved by induction. McCarthy's 91-function was discovered in the context of computer science although it is a purely mathematical result. What is surprising is not the function itself, but the strange induction used to prove it where $98 < 97$. The surprise of the Josephus problem is the bidirectional inductive proof.

## Sources

For a comprehensive presentation of induction see [21]. The proof of McCarthy's 91-function is from [30] where it is attributed to Rod M. Burstall. The presentation of the Josephus problem is based on [21, Chapter 17], which also discusses the historical background. That chapter contains other interesting problems with inductive proofs, such as the muddy children, the counterfeit coin and the pennies in a box. Additional material on the Josephus problem can be found in [44, 57].

# Chapter 7
# Solving Quadratic Equations

Poh-Shen Loh proposed a method for solving quadratic equations that is based on a relation between the coefficients of the quadratic polynomial and its roots. Section 7.1 reviews the traditional methods for solving quadratic equations. Section 7.2 tries to convince the reader that Loh's method makes sense and then explains how to compute the roots. In Sect. 7.3 the computation is carried out for two quadratic polynomials and a similar computation for a quartic polynomial. Section 7.4 derives the traditional formula for the roots from Loh's formulas.

The introduction of algebra and modern algebraic notation is relatively recent. Previously, mathematicians used geometry almost exclusively, so it is interesting to look at al-Khwarizmi's geometric construction of the formula for the roots of quadratic equations (Sect. 7.5). Section 7.6 shows a clever geometric construction used by Cardano in developing the formula for the roots of cubic equations.

Section 7.8 presents other geometric methods for finding the roots of quadratic equations.[1] The chapter concludes with Sect. 7.9 which discusses numerical computation of the roots of quadratic equations.

## 7.1 Traditional Methods for Solving Quadratic Equations

Every student of mathematics memorizes the formula for obtaining the roots of a quadratic equation $ax^2 + bx + c = 0$:

$$x_1, x_2 = \frac{-b \pm \sqrt{b^2 - 4ac}}{2a}.$$

---

[1] Chapter 11 is a prerequisite for a full understanding of these methods.

© The Author(s) 2022
M. Ben-Ari, *Mathematical Surprises*, https://doi.org/10.1007/978-3-031-13566-8_7

For now we will work with monic polynomials, $x^2 + bx + c = 0$, whose roots are:

$$x_1, x_2 = \frac{-b \pm \sqrt{b^2 - 4c}}{2} .$$

(7.1)

Another method of solving quadratic equations is by factoring the polynomials more-or-less by trial-and-error. Sometimes it is easy to obtain the roots by factoring:

$$x^2 - 4x + 3 = (x - 1)(x - 3) .$$

(7.2)

It is much harder to factor $x^2 - 2x - 24$ because there are many possible pairs of roots that must be considered:

$$(\pm 1, \mp 24) , (\pm 2, \mp 12) , (\pm 3, \mp 8) , (\pm 4, \mp 6) .$$

## 7.2 The Relation Between the Roots and the Coefficients

**Theorem 7.1** *If $r_1, r_2$ are the roots of $x^2 + bx + c$ then:*

$$(x - r_1)(x - r_2) = x^2 - (r_1 + r_2)x + r_1 r_2 = x^2 + bx + c .$$

*Therefore, even if we do not know the values of the roots, we do know that:*

$$r_1 + r_2 = -b , \qquad r_1 r_2 = c .$$

(7.3)

There is really nothing to prove because the result emerges from the computation. Consider some values of $-b, r_1, r_2$ and let $m_{12}$ be the average of $r_1, r_2$:

| $-b$ | $r_1$ | $r_2$ | $m_{12}$ |
|------|-------|-------|----------|
| 33 | 12 | 21 | $16\frac{1}{2}$ |
| 33 | 8 | 25 | $16\frac{1}{2}$ |
| 33 | 1 | 32 | $16\frac{1}{2}$ |
| $-b$ | $r_1$ | $r_2$ | $m_{12}$ |
| $-4$ | $-16$ | 12 | $-2$ |
| $-4$ | $-4$ | 0 | $-2$ |
| $-4$ | $-3$ | $-1$ | $-2$ |

For any quadratic equation the average of the two roots is constant:

$$m_{1,2} = \frac{r_1 + r_2}{2} = \frac{(-b - r_2) + r_2}{2} = -\frac{b}{2} .$$

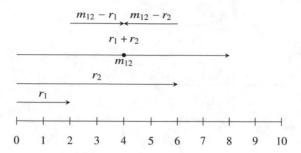

**Fig. 7.1** Relation between the roots $r_1, r_2 = 2, 6$ and their average $m_{12} = 4$

Let $s$ be any number. Then:

$$-b = -b + s + (-s) = \left(\frac{-b}{2} + s\right) + \left(\frac{-b}{2} - s\right) = r_1 + r_2 \,.$$

If one root is at distance $s$ from the average, the other root is at distance $-s$ from the average. For $r_1, r_2 = 2, 6$, where $m_{12} = 4, s = 2$, we have:

| $-b$ | $r_1$ | $r_2$ | $m_{12}$ | $m_{12}-r_1$ | $m_{12}-r_2$ |
|---|---|---|---|---|---|
| 33 | 12 | 21 | $16\frac{1}{2}$ | $4\frac{1}{2}$ | $-4\frac{1}{2}$ |
| 33 | 8 | 25 | $16\frac{1}{2}$ | $8\frac{1}{2}$ | $-8\frac{1}{2}$ |
| 33 | 1 | 32 | $16\frac{1}{2}$ | $15\frac{1}{2}$ | $-15\frac{1}{2}$ |
| $-4$ | $-16$ | 12 | $-2$ | 14 | $-14$ |
| $-4$ | $-4$ | 0 | $-2$ | 2 | $-2$ |
| $-4$ | $-3$ | $-1$ | $-2$ | 1 | $-1$ |

Figure 7.1 visualizes this relation. If we use other values $r_1, r_2 = 3, 5$ for which $r_1 + r_2 = 8$ then $m_{12} = 4$ remains the same while $s$ becomes 1 (Fig. 7.2).

The offset $s$ seems to be arbitrary in:

$$r_1 = \left(\frac{-b}{2} + s\right), \quad r_2 = \left(\frac{-b}{2} - s\right),$$

but there is an additional constraint $r_1 r_2 = c$, where $c$ is the constant term in the polynomial. By multiplying the two expressions we have derived for $r_1, r_2$, we can determine $s$ and then $r_1, r_2$:

$$c = \left(-\frac{b}{2} + s\right)\left(-\frac{b}{2} - s\right) = \frac{b^2}{4} - s^s$$

$$s = \frac{\sqrt{b^2 - 4c}}{2} \,.$$

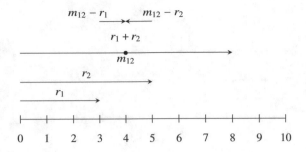

**Fig. 7.2** Relation between the roots $r_1, r_2 = 3, 5$ and their average $m_{12} = 4$

## 7.3 Examples of Loh's Method

*Example 7.1* Consider the polynomial $x^2 - 2x - 24$ where $b = -2, c = -24$:

$$c = \left(-\frac{(-2)}{2} + s\right)\left(-\frac{(-2)}{2} - s\right)$$
$$-24 = (1 + s)(1 - s)$$
$$s = 5$$
$$r_1 = 1 + 5 = 6$$
$$r_2 = 1 - 5 = -4 .$$

Check: $(x - 6)(x - (-4)) = x^2 - 2x - 24$.

*Example 7.2* Let us find the roots of $x^2 - 83x - 2310$:

$$-2310 = \left(\frac{83}{2} + s\right)\left(\frac{83}{2} - s\right)$$
$$s^2 = \frac{6889}{4} + 2310 = \frac{16129}{4}$$
$$s = \frac{127}{2}$$
$$r_1 = \frac{83}{2} - \frac{127}{2} = -22$$
$$r_2 = \frac{83}{2} + \frac{127}{2} = 105 .$$

Check: $(x + 22)(x - 105) = x^2 - 83x - 2310$.

Compare this computation with the computation using the traditional formula:

$$\frac{-b \pm \sqrt{b^2 - 4c}}{2} = \frac{-(-83) \pm \sqrt{(-83)^2 - 4 \cdot (-2310)}}{2}$$

$$= \frac{83 \pm \sqrt{16129}}{2} = \frac{83 \pm 127}{2}$$

$$r_1 = \frac{83 - 127}{2} = -22$$

$$r_2 = \frac{83 + 127}{2} = 105 .$$

*Example 7.3* Theorem 7.1 can be generalized to polynomials of higher degrees. Here is an interesting example for a *quartic equation* $x^4 - 10x^2 - x + 20 = 0$. As with quadratic equations there are formulas for solving cubic and quartic equations (though not equations of higher powers), but the formulas are quite complicated.

Does this polynomial of degree four factor into two quadratic polynomials with integer coefficients? If so, the coefficients of the $x$ terms must be *equal and of opposite signs* since the coefficient of the $x^3$ term is zero. Therefore, the form of the quadratic factors is:

$$f(x) = (x^2 - nx + k_1) (x^2 + nx + k_2) .$$

Carrying out the multiplication results in:

$$f(x) = x^4 + nx^3 + k_2 x^2$$
$$- nx^3 - n^2 x^2 - nk_2 x$$
$$+ k_1 x^2 + nk_1 x + k_1 k_2 .$$

Equating the coefficients gives three equations in the three unknowns $n, k_1, k_2$ gives:

$$(k_1 + k_2) - n^2 = -10$$
$$n(k_1 - k_2) = -1$$
$$k_1 k_2 = 20 .$$

Since we are looking for factors with integer coefficients, from the last two equations it is clear that:

$$n = 1, \; k_1 = 4, \; k_2 = 5 \qquad \text{or} \qquad n = 1, \; k_1 = -5, \; k_2 = -4 .$$

Only $n = 1, k_1 = -5, k_2 = -4$ satisfy the first equation for the coefficient of $x^2$:

$$f(x) = (x^2 - x - 5) (x^2 + x - 4) .$$

Solving these quadratic equations gives four solutions of the quartic equation:

$$x = \frac{1 \pm \sqrt{21}}{2} \quad \text{or} \quad x = \frac{-1 \pm \sqrt{17}}{2} \, .$$

## 7.4 Derivation of the Traditional Formula

For an arbitrary monic polynomial $x^2 + bx + c$, Loh's formulas are:

$$c = r_1 r_2 = \left(\frac{-b}{2} + s\right)\left(\frac{-b}{2} - s\right) = \left(\frac{b^2}{4} - s^2\right)$$

$$s = \sqrt{\left(\frac{b^2}{4}\right) - c}$$

$$r_1, r_2 = \frac{-b}{2} \pm \sqrt{\left(\frac{b^2}{4}\right) - c} = \frac{-b \pm \sqrt{b^2 - 4c}}{2} \, ,$$

the traditional formula for obtaining the roots of a monic quadratic polynomial. If the polynomial is not monic divide it by $a$, substitute in the equation and simplify:

$$x^2 + \frac{b}{a}x + \frac{c}{a} = 0$$

$$r_1, r_2 = \frac{-(b/a) \pm \sqrt{(b/a)^2 - 4(c/a)}}{2}$$

$$= \frac{-b \pm \sqrt{b^2 - 4ac}}{2a} \, .$$

## 7.5 Al-Khwarizmi's Geometric Solution of Quadratic Equations

Let us write a monic quadratic polynomial as $x^2 + bx - c$. The roots can be found by *completing the square*:

$$x^2 + 2\left(\frac{b}{2}\right)x + \left(\frac{b}{2}\right)^2 = c + \left(\frac{b}{2}\right)^2$$

$$\left(x + \frac{b}{2}\right)^2 = c + \left(\frac{b}{2}\right)^2$$

$$x = -\frac{b}{2} \pm \sqrt{c + \left(\frac{b}{2}\right)^2} = \frac{-b \pm \sqrt{b^2 + 4c}}{2} \, .$$

This is the familiar formula for finding the roots of a quadratic equation, except that $4c$ has the opposite sign since the coefficient of the constant term was $-c$.

Completing the square was developed in the 8th century by Muhammad ibn Musa al-Khwarizmi in a geometric context. Given the equation $x^2 + bx = c$, assume that there is a square whose side is $x$ so that its area is $x^2$. To the area $x^2$ add $bx$ by adding four rectangles of area $bx/4$ whose sides are $b/4$ and $x$ (Fig. 7.3a). Now complete the diagram to a square by adding the four little squares of area $(b/4)^2$ (Fig. 7.3b).

We can't construct the diagram in Fig. 7.3a because we don't know what $x$ is, but the area of the larger square in Fig. 7.3b is:

$$x^2 + bx + \frac{b^2}{4} = c + \frac{b^2}{4},$$

which we do know since we are given the coefficients $b, c$. By constructing the diagram and erasing the small squares whose sides are $(b/4)$—another known quantity—we obtain the line segment of length $x$.

*Example 7.4* Let $x^2 + 12x = 64$. Then $c + (b^2/4) = 64 + 36 = 100$. It is easy to construct a square of area 100 since each side has length 10. Now subtract $(b/4) + (b/4) = 6$, the sides of the smaller squares, to get $x = 10 - 6 = 4$.

## 7.6 Cardano's Construction for Solving Cubic Equations

The formula for the roots of cubic equations was first published in the 16th century by Gerolamo Cardano. We will not develop the formula here, but it is interesting that the central idea is based on a geometric construction similar to al-Khwarizmi's.

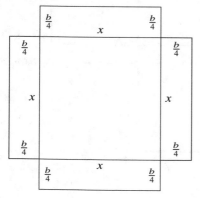

**Fig. 7.3a** The area is $x^2 + 4(b/4)x = x^2 + bx$

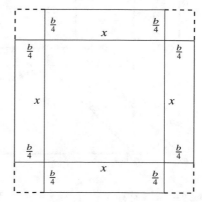

**Fig. 7.3b** The area is $x^2 + 4(b/4)x + 4(b/4)^2 = x^2 + bx + (b^2/4)$

The construction can be obtained very simply using algebra. By multiplication:

$$(a + b)^3 = a^3 + 3a^2b + 3ab^2 + b^3 = (a^3 + b^3) + 3ab(a + b) .\qquad(7.4)$$

Geometrically, we start with a cube whose side is $a + b$ so that its volume is $(a+b)^3$. The cube is decomposed into five pieces. The first two are cubes whose sides are $a$ and $b$ with volumes $a^3$ (blue) and $b^3$ (red), respectively (Fig. 7.4).

The other three parts are boxes (the technical term is *cuboid*) each with one side of length $a + b$ coinciding with a side of the cube, one side of length $a$ and one side of length $b$, so that the volume of each of the three boxes is $ab(a + b)$. In Fig. 7.5, there is one box at the left side of the cube (blue), one at the back of the cube (red) and one at the top of the cube (green). By combining the five solids in Fig. 7.4 and Fig. 7.5 we obtain Eq. 7.4.

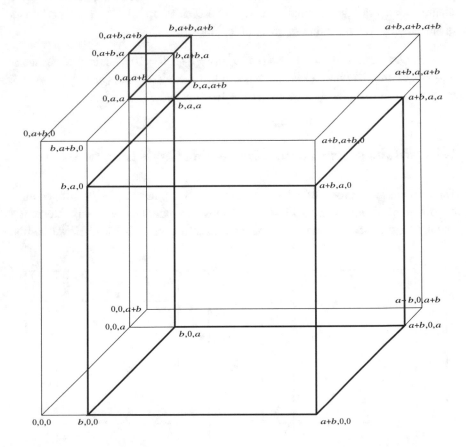

**Fig. 7.4** $(a^3 + b^3) = (a^3 + b^3) + \cdots$

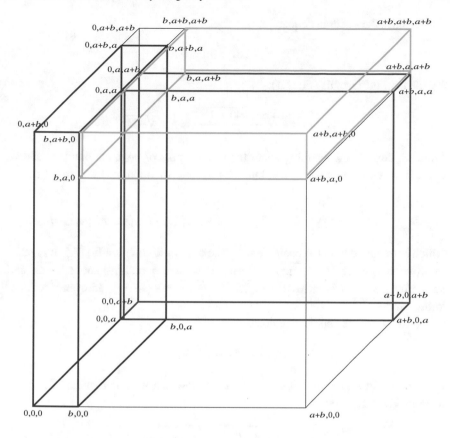

**Fig. 7.5** $(a^3 + b^3) = \cdots + 3ab(a + b)$

## 7.7 They Weren't Intimidated by Imaginary Numbers

The history of mathematics demonstrates a progression of concepts that were initially considered to be meaningless, but were eventually understood, accepted and proved to be useful. "Obviously," since numbers count things, $-1$, a negative number, is meaningless. "Obviously," since numbers are ratios of integers (rational numbers), $\sqrt{2}$, which can easily be proved to be irrational, is meaningless. "Obviously," $\sqrt{-1}$, the square root of a negative number, is meaningless since there is no number—integer, rational or real—whose square is $-1$.

A full understanding of the square roots of negative numbers, to this day called *imaginary numbers* although they are no less real than real numbers, was not achieved until the nineteenth century. Therefore, it is surprising that already in the sixteenth century, Geralamo Cardano and Rafael Bombelli refused to be intimidated by the concept, and took the first small steps towards understanding these numbers.

Consider the quadratic equation:

$$x^2 - 10x + 40 = 0.$$  (7.5)

By the familiar formula (Eq. 7.1):

$$r_1, r_2 = \frac{10 \pm \sqrt{100 - 160}}{2} = 5 \pm \sqrt{-15}.$$

Well, we don't know anything about the square roots of negative numbers and we don't know what these values are, but like Cardano we do know by Thm 7.1 that:

$$r_1 + r_2 = (5 + \sqrt{-15}) + (5 - \sqrt{-15}) = 10 = -b$$
$$r_1 r_2 \;\; = (5 + \sqrt{-15})(5 - \sqrt{-15}) = 25 - 5\sqrt{-15} + 5\sqrt{-15} - (-15) = 40 = c.$$

which correspond with the coefficients of the quadratic equation Eq. 7.5. It is rather intuitive that $\sqrt{-15} + (-\sqrt{-15}) = 0$ even if we know nothing about $\sqrt{-15}$, and, similarly, it is rather intuitive that $\sqrt{-15} \cdot -(\sqrt{-15}) = -(-15) = 15$ even if we don't know what $\sqrt{-15}$ is.

Consider now the cubic equation:

$$x^3 - 15x - 4 = 0.$$  (7.6)

It is not hard to observe that 4 is a root, but how can it be computed? Cardano's formula gives the root:

$$r = \sqrt[3]{2 + 11\sqrt{-1}} + \sqrt[3]{2 - 11\sqrt{-1}},$$  (7.7)

a quite complicated formula that bears no obviously relation to 4.

Bombelli courageously performed the following computation (see Eq. 7.4):

$$(2 + \sqrt{-1})^3 = 8 + 3 \cdot 4\sqrt{-1} + 3 \cdot 2(-1) + (-1\sqrt{-1}) = 2 + 11\sqrt{-1}$$
$$(2 - \sqrt{-1})^3 = 8 - 3 \cdot 4\sqrt{-1} + 3 \cdot 2(-1) - (-1\sqrt{-1}) = 2 - 11\sqrt{-1},$$

and by Eq. 7.7:

$$r = \sqrt[3]{2 + 11\sqrt{-1}} + \sqrt[3]{2 - 11\sqrt{-1}}$$
$$= \sqrt[3]{(2 + \sqrt{-1})^3} + \sqrt[3]{(2 - \sqrt{-1})^3}$$
$$= (2 + \sqrt{-1}) + (2 - \sqrt{-1}) = 4.$$

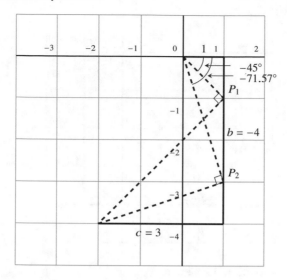

**Fig. 7.6** Lill's method on $x^2 - 4x + 3$

## 7.8 Lill's Method and Carlyle's Circle

Lill's method can be applied to solve quadratic equations.[2] As an example we use
Eq. 7.2 which gives the roots of a quadratic equation obtained by factorization:

$$x^2 + bx + c = x^2 - 4x + 3 = (x - 1)(x - 3) .$$

Applying Lill's method results in the paths shown in Fig. 7.6.
    Check that the angles are correct:

$$- \tan(-45°) = -1, \quad - \tan(-71.57°) \approx -3 .$$

For quadratic equations we can find the points $P_1, P_2$ as the intersections of the line
representing the coefficient $b$ and the circle whose diameter is the line connecting
the starting point and the end point of the paths (Fig. 7.7). In order for a point on the
line $b$ to be a root, the reflection of the line must be 90° and therefore the inscribed
angle is subtended by a diameter.
    This can also be checked by computation. The center of the circle is the midpoint
of the diameter $(-1, -2)$. The length of the diameter is:

$$\sqrt{(-2)^2 + (-4)^2} = \sqrt{20} ,$$

_____

[2] This section assumes that you have read about Lill's method in Chap. 11.

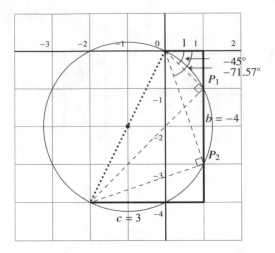

**Fig. 7.7** Constructing a circle to find the roots

so the square of the length of the radius is $\left(\sqrt{20/2}\right)^2 = 5$. We need the intersection of this circle and the line $x = 1$:

$$(x - (-1))^2 + (y - (-2))^2 = r^2$$
$$(x^2 + 2x + 1) + (y^2 + 4y + 4) = 5$$
$$y^2 + 4y + 3 = 0$$
$$y = -1, \ -3 \,.$$

A similar method for solving quadratic equations is the Carlyle circle which predates Lill's method. Given a quadratic equation $x^2 - bx + c$ (note the minus sign on the linear term), construct points at $(0, 1)$ and $(b, c)$. Construct a circle whose diameter is the line connecting the two points (Fig. 7.8). Its intersections (if any) with the $x$-axis are the roots of the equation.

In the general case, the center of the circle is $(b/2, (c - (-1))/2)$ and the length of the diameter is $\sqrt{b^2 + (c - 1)^2}$, so the equation of the circle is:

$$\left(x - \frac{b}{2}\right)^2 + \left(y - \frac{c+1}{2}\right)^2 = \frac{b^2 + (c - 1)^2}{4} \,.$$

For the example, substituting $b = 4$, $c = 3$ and $y = 0$, we see that $x = 1$ and $x = 3$ are the roots of the quadratic equation.

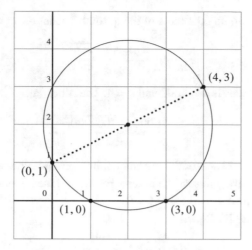

**Fig. 7.8** Carlyle circle for $x^2 - 4x + 3$

## 7.9 Numerical Computation of the Roots

Students learn symbolic computation of roots, derivatives and so on. Today, most computation is performed by computers so symbolic computation is less important. *Numerical analysis* is the branch of mathematics and computer science that develops accurate and efficient computational methods. The main challenge is to deal with the finiteness of values stored in the computer's memory. The computation:

$$0.12 \times 0.14 = 0.0168$$

is easy to do, but:

$$0.123456789 \times 0.123456789$$

needs eighteen digits to be accurately represented and this cannot be done in a memory word that stores sixteen digits. This error is called a *round-off error*.

An even more serious problem is encountered when *floating point arithmetic* is performed. Clearly:

$$(0.12 \times 10^{-10}) \times (0.14 \times 10^{-8})$$

would not be computed by writing out all the zero digits. Instead, we multiply the mantissas and add the exponents to obtain $0.0168 \times 10^{-18}$, which is normalized to $0.168 \times 10^{-19}$ so that the most significant digit appears after the decimal point, ensuring maximum precision given the fixed size of the mantissa. If the maximum exponent that can be represented is $-16$ the result simply cannot even be stored. This error is called *floating-point underflow*.

The formula for finding the roots of the quadratic equation $x^2 + bx + c$ is:

$$r_1, r_2 = \frac{-b \pm \sqrt{b^2 - 4c}}{2} . \tag{7.8}$$

Consider what happens if $b = 1000$ and $c = 4$. The roots are:

$$r_1, r_2 = \frac{-1000 \pm \sqrt{1000000 - 16}}{2} .$$

Depending on the precision of the arithmetic, it is possible that one of the roots is so close to zero that the value stored is zero. Evaluating the quadratic equation gives the surprising result $0^2 + b \cdot 0 + 4 = 4 = 0$.

Can we do better? By Eq. 7.3:

$$r_1 + r_2 = -b , \qquad r_1 r_2 = c .$$

If $r_2$ is much less that $r_1$, written $r_2 \ll r_1$, then $r_1 \approx -b$ and $r_2 = c/b$. Table 7.1, computed by a computer program, compares the values of the roots computed by these formulas with the values obtained from the traditional formula Eq. 7.8. The value of $c$ is fixed at 4 and the roots for increasing values of $b$ are shown.

Initially, the true values computed by the traditional formula for $r_2$ are more accurate ($r_2 - r_{2v}$ is negative) but from $b = 100000$, the computation based upon Eq. 7.3 is more accurate. Such are the surprises of numerical analysis.

**Table 7.1** Two computations of the roots of a quadratic equation. $r_1, r_2$ are the roots computed by Eq. 7.8. $r_{1v}, r_{2v}$ are the roots computed using Eq. 7.3. The errors are $r_i - r_{iv}$. The values are truncated to four decimal places. Floating-point numbers are written $-4e - 5$ in place of $4 \times 10^{-5}$ because computer programs are normally written as linear sequences of characters.

| $b$ | $r_1$ | $r_{1v}$ | Error$_1$ | $r_2$ | $r_{2v}$ | Error$_2$ |
|---|---|---|---|---|---|---|
| 100 | $-99.9599$ | $-100$ | $0.0400$ | $-0.04001$ | $-0.04$ | $-1.6012e-05$ |
| 1000 | $-999.9959$ | $-1000$ | $0.0040$ | $-0.0040$ | $-0.004$ | $-1.6000e-08$ |
| 10000 | $-9999.9996$ | $-10000$ | $0.0004$ | $-0.0004$ | $-0.0004$ | $-1.6270e-11$ |
| 100000 | $-99999.9999$ | $-100000$ | $3.9999e-5$ | $-3.9999e-5$ | $-4e-5$ | $1.0104e-12$ |
| 1000000 | $-999999.9999$ | $-1000000$ | $4.0000e-6$ | $-3.9999e-6$ | $-4e-6$ | $2.7749e-11$ |
| 10000000 | $-10000000.0$ | $-10000000$ | $3.9860e-7$ | $-3.9953e-7$ | $-4e-7$ | $4.6261e-10$ |

## What Is the Surprise?

Poh-Shen Loh's approach provides a new way of looking at the relation between the coefficients and the roots that one doesn't see simply by memorizing the traditional formula. What is surprising is that this relation is fundamental in Gauss's algebraic proof of the constructibility of a regular heptadecagon (Chap. 16).

With the modern dominance of algebraic methods in geometry it is important to be reminded that the reverse once held. As shown by the constructions of Al-Khwarizmi and Cardano, geometric methods were used to obtain results in algebra. Lill and Carlyle both developed geometric methods for solving quadratic equations. Considerations of numerical computation on computers will surprise students who have not experienced it before.

## Sources

Poh-Shen Loh's method is from [28, 29]. Al-Khwarizmi's construction is from [6, Chapter 1] and [32]. Cardano's construction can be found in [6, Chap. 1]. For the colorful history of the development of Cardano's formula see [52]. The early attempts at computing with imaginary numbers are from [6, Chapter 2]. Lill's method and Carlyle's circle can be found in [61] together with a discussion of numerical computation of the roots.

# Chapter 8
# Ramsey Theory

Ramsey theory is a branch of combinatorics that asks questions of the form: How large must a set be so that if it is divided into subsets, at least one subset has a certain property? Results in Ramsey theory are difficult to prove and there remain many open problems. In this chapter we present simple cases of four problems to give a taste of this fascinating subject: Schur triples (Sect. 8.1)—triples of integers such that $a + b = c$, Pythagorean triples (Sect. 8.2)—triples of integers such that $a^2 + b^2 = c^2$, van der Waarden's problem (Sect. 8.3) which concerns sequences of numbers, and Ramsey's theorem (Sect. 8.4) on coloring graphs. Section 8.5 shows how the probabilistic method in combinatorics can be used to develop a lower bound for Ramsey numbers.

The Pythagorean triples problem was recently solved with the aid of computers, using a relatively new method call SAT solving. For readers familiar with propositional logic Sect. 8.6 gives an overview of how this is done.

Section 8.7 describes Pythagorean triples as known to the Babylonians four thousand years ago.

Terminology: *monochromatic* means *of the same color*.

## 8.1 Schur triples

**Definition 8.1** Given *any* decomposition of the set of positive integers:

$$S(n) = \{1, \ldots, n\}$$

into two disjoint subsets $S_1, S_2$, do there exist $\{a, b, c\} \subseteq S_1$ or $\{a, b, c\} \subseteq S_2$ (or both) such that $a < b < c$ and $a + b = c$? If so, the set $\{a, b, c\}$ is called a *Schur triple*.

*Example 8.1* For $n = 8$, in the decomposition:

$$S_1 = \{1, 2, 3, 4\}, \ S_2 = \{5, 6, 7, 8\}, \tag{8.1}$$

© The Author(s) 2022
M. Ben-Ari, *Mathematical Surprises*, https://doi.org/10.1007/978-3-031-13566-8_8

the set $S_1$ includes the Schur triple $\{1, 2, 3\}$. However, the decomposition:

$$S_1' = \{1, 2, 4, 8\}, \quad S_2' = \{3, 5, 6, 7\}, \tag{8.2}$$

does not contain a Schur triple, as you can check by enumerating all the triples in each subset.

**Theorem 8.1** *In all* decompositions of $S(9) = \{1, \ldots, 9\}$ *into two disjoint subsets, at least one subset contains a Schur triple.*

Of course we could check the $2^9 = 512$ decompositions of $S(9)$ into two disjoint subsets, but let us try come up with a more succinct proof.

**Proof** We try to construct a decomposition that *does not* contain a Schur triple and show that the constraints of the problem make this impossible. Start by placing 1 and 3 into the subset $S_1$. 2 must be placed in $S_2$ because $1 + 2 = 3$ and we are trying to construct a decomposition that does not contain a Schur triple. Similarly, 4 must be placed in $S_2$ because $1 + 3 = 4$. Continuing, 6 is placed in $S_1$ because $2 + 4 = 6$ and 7 is placed in $S_2$ because $1 + 6 = 7$. However, $3 + 6 = 9$ and $2 + 7 = 9$, so 9 must appear in both $S_1$ and $S_2$, a contradiction. The sequence of inferences is shown in the following table:

| $S_1$ | $S_2$ |
|---|---|
| 1, 3 | |
| 1, 3 | 2 |
| 1, 3 | 2, 4 |
| 1, 3, 6 | 2, 4 |
| 1, 3, 6 | 2, 4, 7 |
| 1, 3, 6, 9 | 2, 4, 7 |
| 1, 3, 6, 9 | 2, 4, 7, 9 |

Backtracking, we search for a decomposition where 1, 3 are in different subsets. If we place 5 into $S_2$, a sequence of inferences again leads to a contradiction because 9 must appear in both subsets. The reader should justify each of the inferences shown in the following table:

| $S_1$ | $S_2$ |
|---|---|
| 1 | 3 |
| 1 | 3, 5 |
| 1, 2 | 3, 5 |
| 1, 2, 8 | 3, 5 |
| 1, 2, 8 | 3, 5, 7 |
| 1, 2, 8 | 3, 5, 7, 9 |
| 1, 2, 8 | 3, 5, 6, 7, 9 |
| 1, 2, 8, 9 | 3, 5, 6, 7, 9 |

Backtracking again, we try to place 5 into $S_1$, but that also leads to a contradiction, as shown in the following table:

| $S_1$ | $S_2$ |
|---|---|
| 1 | 3 |
| 1, 5 | 3 |
| 1, 5 | 3, 4 |
| 1, 5 | 3, 4, 6 |
| 1, 2, 5 | 3, 4, 6 |
| 1, 2, 5 | 3, 4, 6, 7 |
| 1, 2, 5, 7 | 3, 4, 6, 7 |

If follows that there is no decomposition that does not include a Schur triple.   □

Issai Schur proved the following theorem:

**Theorem 8.2 (Schur)** *For every $k \geq 2$ there is a smallest $n$ such that in any disjoint decomposition of $S(n)$ into $k$ subsets, at least one of the subsets must contain a Schur triple.*

## 8.2 Pythagorean Triples

**Definition 8.2** Given *any* decomposition of the set of positive integers:

$$S(n) = \{1, \ldots, n\}$$

into two disjoint subsets $S_1, S_2$, do there exist $\{a, b, c\} \subseteq S_1$ or $\{a, b, c\} \subseteq S_2$ (or both) such that $a < b < c$ and $a^2 + b^2 = c^2$? If so, $\{a, b, c\}$ is called a *Pythagorean triple*.

*Example 8.2* For $n = 10$, in the decomposition into even and odd numbers:

$$S_1 = \{1, 3, 5, 7, 9\}, \quad S_2 = \{2, 4, 6, 8, 10\},$$

there are no Pythagorean triples in $S_1$ but $\{6, 8, 10\}$ in $S_2$ is a Pythagorean triple since $6^2 + 8^2 = 10^2$.

Marijn J.H. Heule and Oliver Kullmann proved the following theorems. Their method of proof is discussed in Sect. 8.6.

**Theorem 8.3** *For all $n \leq 7824$, there is some decomposition of $S(n)$ into two disjoint subsets such that both subsets do not contain a Pythagorean triple.*

**Theorem 8.4** *For all $n \geq 7825$, in* all *decompositions of $S(n)$ into two disjoint subsets at least one subset* contains *a Pythagorean triple.*

It is impossible to check all $2^{7825}$ decompositions of $S(7825)$. If we could check one decomposition every microsecond, $2^{7825}$ microseconds $\approx 10^{600}$ years, while the estimated age of the universe is only about $10^{10}$ years.

## 8.3 Van der Waerden's problem

Consider the sequences of eight colored dots in Fig. 8.1. In the top sequence there are red dots at positions $(1, 2, 3)$ and blue dots at positions $(4, 5, 6)$. In each case, the positions form an arithmetic progression. Similarly, in the middle sequence the red dots at positions $(1, 3, 5)$ form an arithmetic progression. However, in the bottom sequence there is no set of three monochromatic dots whose positions form an arithmetic progression. Triples of red dots are at positions $(1, 2, 5)$, $(1, 2, 6)$, $(2, 5, 6)$, none of which form arithmetic progressions, and similarly for the blue dots.

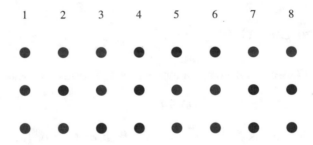

**Fig. 8.1** van der Waerden's problem for eight colored dots

With nine dots *any* coloring *must* contain a sequence of three monochromatic dots that form an arithmetic progression. For example, let us add a red dot or a blue dot at the end of the bottom sequence in Fig. 8.1 to obtain the sequences in Fig. 8.2. In the top sequence there are red dots at positions $(1, 5, 9)$, an arithmetic progression, and in the bottom sequence there are blue dots at positions $(7, 8, 9)$, also an arithmetic progression.

Bartel L. van der Waerden posed the following problem: For any positive integer $k$, what is the smallest number $n$ such that *any* sequence of $n$ colored dots *must* contain a sequence of $k$ monochromatic dots that form an arithmetic progression? For $k = 3$, $n = 9$, as demonstrated above for one decomposition. The next result is more difficult to show: for $k = 4$, $n = 35$.

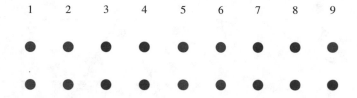

1  2  3  4  5  6  7  8  9

**Fig. 8.2** van der Waerden's problem for nine colored dots

## 8.4 Ramsey's Theorem

Color the edges of $K_5$, the complete graph on 5 vertices, with two colors as shown in Fig. 8.3a. There are no monochromatic subgraphs $K_3$ (triangles) in the graph. Figure 8.3b shows one coloring of $K_6$ and it is easy to see that there are monochromatic triangles $\triangle ACE$ and $\triangle BDF$. In this section we prove a simple case of a theorem by Frank P. Ramsey on the existence of subsets with a certain property.

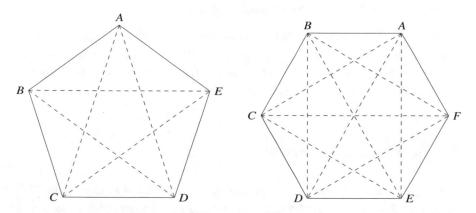

**Fig. 8.3a** A coloring of $K_5$ with two colors     **Fig. 8.3b** A coloring $K_6$ with two colors

**Definition 8.3**     $R(k)$, the *Ramsey number* for $k$, is the smallest number $n$ such that in *any* coloring of $K_n$, the complete graph on $n$ vertices, with two colors there is a monochromatic complete subgraph $K_k$.

**Theorem 8.5 (Ramsey)**     $R(3) = 6$.

**Proof** Figure 8.3a shows that $R(3) > 5$. To show that $R(3) \leq 6$, consider any vertex $v$ in $K_6$. $v$ is connected to five other vertices, and when the edges are colored with two colors there must be at least three monochromatic edges incident with $v$.

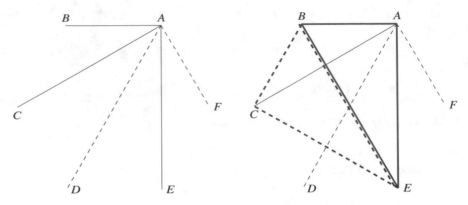

**Fig. 8.4a** One vertex of $K_6$          **Fig. 8.4b** Monochromatic triangles in $K_6$

In Fig. 8.4a, $\overline{AB}, \overline{AC}, \overline{AE}$ are colored red. Since the graph is complete all the vertices are connected, so if any one of the edges $\overline{BC}, \overline{BE}, \overline{CE}$ is colored red, say $\overline{BE}$, a red triangle $\triangle ABE$ is formed. Otherwise, all three edges of these edges are colored blue and they form a blue triangle (Fig. 8.4b).                    □

The theorem can be generalized to any number of colors, as well as to colorings where the sizes of the subgraphs are not the same. $R(r, b, g)$ is the smallest complete graph such that in any coloring with three colors there must be complete subgraphs with $r$ red edges, $b$ blue edges and $g$ green edges.

## 8.5 The Probabilistic Method

The only known non-trivial Ramsey numbers are $R(3) = 6$ and $R(4) = 18$. In 1947 Paul Erdős developed the *probabilistic method* and used it to show lower and upper bounds on $R(k)$. Subsequent research has improved both bounds, but this is still a significant research area since the bounds are not tight. For example, it has been proved that $43 \leq R(5) \leq 48$ and $798 \leq R(10) \leq 23556$. In this section elementary probability is used to obtain a lower bound on $R(k)$.

To show that there exists an element of a set $S$ that has property $A$, prove that the probability of a *random* element of $S$ having property $A$ is greater than zero. It is important to understand that the method is *non-constructive*: it just proves that such an element exists but does not construct one. Although from Thm. 8.5 we know that $R(3) = 6$, let us use the probabilistic method to obtain a lower bound for $R(3)$.

**Theorem 8.6 (Erdős)** $R(3) > 4$.

**Proof** Given a *random* coloring of $K_n$ by the two colors red and blue, consider an arbitrary subgraph $K_3$, that is, an arbitrary triangle with $\binom{3}{2} = 3$ sides. The probability that all sides are colored red is $2^{-3}$, as is the probability that all sides are colored blue, so the probability that the triangle is monochromatic is $2^{-3} + 2^{-3} = 2^{-2} = 1/4$. The number of triangles in $K_n$ is $\binom{n}{3}$, so $P(n, 3)$, the probability that *some* triangle contained in a random coloring of $K_n$ is monochromatic, is:

$$P(n, 3) = \binom{n}{3} \cdot \frac{1}{4}.$$

If $P(n, 3) < 1$ then its complement $\overline{P}(n, 3) = 1 - P > 0$, that is, the probability that a random coloring of $K_n$ does *not* contain a monochromatic triangle is greater than zero, so at least one must exist.

The following table shows $\overline{P}(n, 3)$ for several values of $n$, and whether the value of $\overline{P}(n, 3)$ proves that there exists a coloring with no monochromatic triangle:

| $n$ | $\overline{P}(n, 3)$ | *Exists* |
|---|---|---|
| 3 | 3/4 | *yes* |
| 4 | 5/6 | *yes* |
| 5 | −3/7 | − |

□

At first glance the result is strange because Fig. 8.3a shows that there exists a coloring of $K_5$ with no monochromatic coloring. However, the probabilistic criterion is sufficient but not necessary; it is a lower bound, meaning that $R(n) > 4$ which is true because Thm. 8.5 showed that $R(n) = 6$.

The same proof works for arbitrary $k$, so the probability of the existence of a coloring of $K_n$ with no monochromatic complete graph $K_k$ is:

$$P(n, k) = \binom{n}{k} \cdot 2 \cdot 2^{-\binom{k}{2}}.$$

For $k = 4$:

$$\overline{P}(n, 4) = 1 - \binom{n}{4} \cdot 2^{-5} = \left(32 - \binom{n}{4}\right) \bigg/ 32$$

$$\overline{P}(6, 4) = (32 - 15)/32 = 17/32$$

$$\overline{P}(7, 4) = (32 - 35)/32 = -3/32.$$

If follows that $R(4) > 6$ which is much less than the known value $R(4) = 18$.

## 8.6 SAT Solving

SAT solving is a method for solving problems that works by encoding a problem as a formula in propositional logic and then using a computer program to check the truth value of the formula. Advances in algorithms and implementations have made SAT solving a viable approach for problem solving. We give an overview of SAT solving and explain how it can be used to solve the mathematical problems described in this chapter. The reader is assumed to have an elementary knowledge of propositional logic as summarized in Def. 8.4.

### 8.6.1 Propositional Logic and the SAT Problem

**Definition 8.4**

- A *formula* is composed of *atomic formulas* or *atoms* connected by the propositional operators $\vee$ (disjunction, "or"), $\wedge$ (conjunction, "and"), $\neg$ (negation, "not").
- A formula is given an *interpretation* by an assignment of $T$ or $F$ to each atom. Evaluating a formula in an interpretation results in its *truth value $T$ or $F$*.
- A formula is *satisfiable* if and only if there is an interpretation that makes its truth value $T$. Otherwise, the formula is *unsatisfiable*.
- A formula is in *conjunctive normal form (CNF)* if and only if it is composed of a conjunction of subformulas each of which is a disjunction of *literals* (atoms or negations of atoms).

  The following formula is in CNF:

  $$(\neg p \vee q \vee \neg r) \wedge (\neg p \vee r) \wedge (\neg r) \wedge (p \vee q \vee \neg r).$$

The *SAT problem* is to decide if a given formula in CNF is satisfiable or not. A *SAT solver* is a computer program that can solve the SAT problem. Most SAT solvers are based on the DPLL algorithm which goes back to the 1960's, but recent developments have made very significant improvements to the algorithm. Highly optimized implementations of these algorithms have made SAT solvers an important tool for solving problems in many fields including mathematics.

## 8.6.2 Schur triples

Let us encode the Schur triples problem $S(8)$ as a formula in CNF. The formula will
be satisfiable if and only if there is a decomposition of a set $S$ into disjoint subsets
$S_1, S_2$ such that neither $S_1$ nor $S_2$ contains a Schur triple. There is an atom $p_i$ for
each of the numbers $1 \leq i \leq 8$. The intended meaning of an interpretation for the
formula is that it assigns $T$ to $p_i$ if $i$ is in the first subset $S_1$ and it assigns $F$ to $p_i$ if $i$
is in the second subset $S_2$. To show that in all decompositions neither subset contains
a Schur triple, the interpretation must ensure that for each possible Schur triple at
least one atom is assigned $T$ and one atom is assigned $F$.

For example, $\{2, 4, 6\}$ is a Schur triple so at least one of the three integers must
be in $S_1$ and at least one of them must be in $S_2$. Therefore, $p_2 \vee p_4 \vee p_6$ must be true
and also $\neg p_2 \vee \neg p_4 \vee \neg p_6$ must be true. There are 12 possible Schur triples so the
CNF formula is:

$$
\begin{aligned}
&(p_1 \vee p_2 \vee p_3) \wedge (\neg p_1 \vee \neg p_2 \vee \neg p_3) \wedge \\
&(p_1 \vee p_3 \vee p_4) \wedge (\neg p_1 \vee \neg p_3 \vee \neg p_4) \wedge \\
&(p_1 \vee p_4 \vee p_5) \wedge (\neg p_1 \vee \neg p_4 \vee \neg p_5) \wedge \\
&(p_1 \vee p_5 \vee p_6) \wedge (\neg p_1 \vee \neg p_5 \vee \neg p_6) \wedge \\
&(p_1 \vee p_6 \vee p_7) \wedge (\neg p_1 \vee \neg p_6 \vee \neg p_7) \wedge \\
&(p_1 \vee p_7 \vee p_8) \wedge (\neg p_1 \vee \neg p_7 \vee \neg p_8) \wedge \\
&(p_2 \vee p_3 \vee p_5) \wedge (\neg p_2 \vee \neg p_3 \vee \neg p_5) \wedge \\
&(p_2 \vee p_4 \vee p_6) \wedge (\neg p_2 \vee \neg p_4 \vee \neg p_6) \wedge \\
&(p_2 \vee p_5 \vee p_7) \wedge (\neg p_2 \vee \neg p_5 \vee \neg p_7) \wedge \\
&(p_2 \vee p_6 \vee p_8) \wedge (\neg p_2 \vee \neg p_6 \vee \neg p_8) \wedge \\
&(p_3 \vee p_4 \vee p_7) \wedge (\neg p_3 \vee \neg p_4 \vee \neg p_7) \wedge \\
&(p_3 \vee p_5 \vee p_8) \wedge (\neg p_3 \vee \neg p_5 \vee \neg p_8) .
\end{aligned}
\tag{8.3}
$$

When a SAT solver is given this formula it answers that the formula is satisfiable
under either of the interpretations:

| $p_1$ | $p_2$ | $p_3$ | $p_4$ | $p_5$ | $p_6$ | $p_7$ | $p_8$ |
|---|---|---|---|---|---|---|---|
| F | F | T | F | T | T | T | F |
| T | T | F | T | F | F | F | T |

One interpretation corresponds to the decomposition in Eq. 8.2: $S_1 = \{1, 2, 4, 8\}$,
$S_2 = \{3, 5, 6, 7\}$, while the other corresponds to the symmetrical decomposition
$S_1 = \{3, 5, 6, 7\}$, $S_2 = \{1, 2, 4, 8\}$.

For $S(9)$, four pairs of subformulas are added for the additional possible triples:

$$(p_1 \lor p_8 \lor p_9) \land (\neg p_1 \lor \neg p_8 \lor \neg p_9) \land$$
$$(p_2 \lor p_7 \lor p_9) \land (\neg p_2 \lor \neg p_7 \lor \neg p_9) \land$$
$$(p_3 \lor p_6 \lor p_9) \land (\neg p_3 \lor \neg p_6 \lor \neg p_9) \land$$
$$(p_4 \lor p_5 \lor p_9) \land (\neg p_4 \lor \neg p_5 \lor \neg p_9).$$

When the SAT solver is given this formula, it answers that the formula is unsatisfiable, meaning that *no* decomposition has *no* Schur triple. Removing the double negative, this states that in *every* decomposition of $S(9)$ there exists a Schur triple.

### 8.6.3 Pythagorean Triples

Heule and Kullmann solved the Pythagorean triple problem using a highly optimized SAT solver. There was a significant difference in efficiency between finding a decomposition that does not have Pythagorean triples (you just need one decomposition), and showing all that decompositions have a Pythagorean triple (you have to check all of them). To show that for all $S(n)$, $1 \leq n \leq 7824$, there is a decomposition with no triple took only one minute of computing time, whereas to show that every decomposition of $S(7825)$ has a triple took about two days of computing time for a computer with 800 *cores* (processors) working in parallel, altogether $40,000$ hours of computing time.

The use of computers in mathematics naturally raises the question: Can we trust a proof generated by a computer? Of course, even "ordinary" mathematical proofs can be incorrect (Sect. 4.7), but our experience with frequent computer bugs, as well as the opaqueness of large computer programs, makes us more sensitive to potential errors in computer-generated proofs.

One approach to increasing confidence in the correctness of a computer-generated proof is to use two or more programs, written independently by two or more researchers. If the multiple programs are written in different programming languages and for different computers and operating systems, this lessens the possibility of a bug in the computer hardware and software.

Heule and Kullmann's SAT solver wrote out a log of the steps in the proof so that it could be examined for correctness. The log was so massive, 200 terabytes, that it was impossible to examine directly. To put this into perspective, 200 terabytes is 200,000 gigabytes while your computer might have an internal memory of 16 gigabytes and a solid-state disk of 128 gigabytes. Instead, they wrote a small program to verify the correctness of the data in the log. To ensure that *this* program was correct, they wrote a formal proof using the Coq proof assistant that supports and checks the work of mathematicians without totally automating the proof process.

## 8.6.4 An Overview of the DPLL Algorithm

The first algorithm that one learns for SAT solving is *truth tables*. Given a formula $A$ in propositional logic with $n$ different atoms, there are $2^n$ interpretations since each atom can be independently assigned $T$ or $F$. For each interpretation it is straightforward to compute the truth value of $A$ using the definition of the propositional operators. However, to check $2^n$ interpretations is very inefficient for even moderately large $n$.

The DPLL algorithm works by incrementally assigning $T$ or $F$ to an atom and then attempting to evaluate the formula. For example, given $A = p \wedge q \wedge \neg r$, if $p$ is assigned $F$ then $A$ evaluates to $F$, regardless of the assignments to $q$ and $r$, and there is no need to perform further evaluations. Similarly, $A = p \vee q \vee \neg r$ evaluates to $T$ if $p$ is assigned $T$, regardless of the assignments to $q$ and $r$.

The efficiency of DPLL comes from *unit propagation*. Consider part of the formula for Schur triples:

$$(p_1 \vee p_2 \vee p_3) \wedge (\neg p_1 \vee \neg p_2 \vee \neg p_3) \wedge$$
$$(p_1 \vee p_3 \vee p_4) \wedge (\neg p_1 \vee \neg p_3 \vee \neg p_4) \wedge$$
$$\cdots \tag{8.4}$$
$$(p_3 \vee p_4 \vee p_7) \wedge (\neg p_3 \vee \neg p_4 \vee \neg p_7) \wedge$$
$$(p_3 \vee p_5 \vee p_8) \wedge (\neg p_3 \vee \neg p_5 \vee \neg p_8) \,.$$

Suppose that we have assigned $F$ to $p_1, p_2$. The first subformula reduces to the unit formula consisting of the single atom $p_3$. If the formula is to be satisfied, we *must* assign $T$ to $p_3$ and all the subformulas:

$$(p_1 \vee p_2 \vee p_3), \ (p_1 \vee p_3 \vee p_4), \ (p_3 \vee p_4 \vee p_7), \ (p_3 \vee p_5 \vee p_8),$$

immediately evaluate to $T$.

Since $\neg p_3$ evaluates to $F$, each subformula containing $\neg p_3$ can be satisfied only if some other literal in the subformula is assigned $T$. In $\neg p_3 \vee \neg p_5 \vee \neg p_8$, either $p_5$ or $p_8$ must be assigned $F$ so that either $\neg p_5$ or $\neg p_8$ evaluates to $T$.

This analysis shows that once $p_1, p_2$ have been assigned $F$, the formula in Eq. 8.4 is satisfiable if and only if $(\neg p_4 \vee \neg p_7) \wedge (\neg p_5 \vee \neg p_8)$ is satisfiable. By performing the propagation of $p_3$ on all the subformulas of Eq. 8.3, the formula is reduced to:

$$(p_4 \vee p_5) \wedge (p_4 \vee p_6) \wedge (p_5 \vee p_6) \wedge (p_5 \vee p_7) \wedge$$
$$(p_6 \vee p_7) \wedge (p_6 \vee p_8) \wedge (p_7 \vee p_8) \wedge$$
$$(\neg p_4 \vee \neg p_7) \wedge (\neg p_5 \vee \neg p_8) \,.$$

One more assignment of $F$ to $p_4$ results in a satisfying interpretation which we have found after only three arbitrary assignments.

## 8.7 Pythagorean Triples in Babylonian Mathematics

This section is a digression from Ramsey theory; it is included to give a taste of the rich theory of Pythagorean triples and to demonstrate the depth of mathematical knowledge in the ancient world. Pythagorean triples were known in Babylonian mathematics since at least 1800 BCE.

**Definition 8.5** A *primitive Pythagorean triple* is a set of three positive integers $\{a, b, c\}$ such that $a^2 + b^2 = c^2$ and $a, b, c$ have no common factor greater than 1.

*Example 8.3* $\{3, 4, 5\}$ is a primitive Pythagorean triple but $\{6, 8, 10\}$ is a Pythagorean triple that is not primitive since 2 is a common factor.

A cuneiform tablet called *Plimpton* 322 is one of the earliest examples of Babylonian mathematics. It lists fifteen primitive Pythagorean triples by giving $a$ and $c$. Table 8.1 displays four of these triples, together with the computed values of $b$ and other values that will be discussed below. Historians of mathematics have proposed several explanations for how these triples were found. One explanation is that *Euclid's formula* was used to obtain the triples from a pair of generating numbers.

**Theorem 8.7 (Euclid)** $\{a, b, c\}$ *is primitive Pythagorean triple if and only if there exist two positive integers* $u, v,$ *called* generating numbers, *such that:*

*1. $u > v$*
*2. they are not both odd*
*3. they have no common factor greater than 1*
*4. the following relations hold between $\{a, b, c\}$ and $u, v$:*

$$a = u^2 - v^2, \quad b = 2uv, \quad c = u^2 + v^2.$$

**Proof** By computation it follows immediately that if $\{a, b, c\}$ can be expressed as required in item 4 they form a Pythagorean triple:

**Table 8.1** Babylonian triples from the Plimpton 322 tablet

| $a$ | $a_{factors}$ | $b$ | $b_{factors}$ | $c$ | $u$ | $u_{factors}$ | $v$ | $v_{factors}$ |
|---|---|---|---|---|---|---|---|---|
| 119 | $7 \cdot 17$ | 120 | $2^3 \cdot 3 \cdot 5$ | 169 | 12 | $2^2 \cdot 3$ | 5 | 5 |
| 4601 | $43 \cdot 107$ | 4800 | $2^6 \cdot 3 \cdot 5^2$ | 6649 | 75 | $3 \cdot 5^2$ | 32 | $2^5$ |
| 12709 | $71 \cdot 179$ | 13500 | $2^2 \cdot 3^3 \cdot 5^3$ | 18541 | 125 | $5^3$ | 54 | $2 \cdot 3^3$ |
| 65 | $5 \cdot 13$ | 72 | $2^3 \cdot 3^2$ | 97 | 9 | $3^2$ | 4 | $2^2$ |

$$a^2 + b^2 = (u^2 - v^2)^2 + (2uv)^2$$
$$= u^4 - 2(uv)^2 + v^4 + 4(uv)^2$$
$$= u^4 + 2(uv)^2 + v^4$$
$$= u^2 + v^2 = c^2 .$$

The proof of the other direction is more complicated and is omitted. □

If it is true that the Babylonians used Euclid's formula, the question remains: How did they discover the generating numbers $u, v$?

Each row of Table 8.1 displays $a_{factors}$ and $b_{factors}$, the factorizations of $a$ and $b$, respectively, showing that they have no common factors. The reader is invited to check that $c$ has no common factor with $a, b$ so the triples are primitive. The generating numbers $u, v$ and $u_{factors}, v_{factors}$ are also displayed. Not only do they not have any common factors as required by Thm. 8.7, but the only factors greater than 1 in $u$ and $v$ are powers of $2, 3, 5$.

**Definition 8.6** A *Babylonian triple* is a primitive Pythagorean triple such that the only prime factors of $u, v$ are $2, 3, 5$.

The reason that the Babylonians restricted themselves to these factors is that they used the *sexagesimal* or base $60 = 2 \cdot 2 \cdot 3 \cdot 5$ number system whose prime factors are 2, 3 and 5.

For readers who are not familiar with non-decimal number systems, here is a brief overview of the concept. The "number" 12345 is a shorthand for the number:

$$(1 \times 10^4) + (2 \times 10^3) + (3 \times 10^2) + (4 \times 10^1) + (5 \times 10^0) .$$

This number system is called the *decimal* or base 10 number system. There are ten digits $0, 1, 2, \ldots, 8, 9$ for the coefficients of the powers, and the powers are represented by the places of coefficients with powers increasing from right to left.

The number could also be represented in the binary or base 2 number system by:

$$12345 = 8192 + 4096 + 32 + 16 + 8 + 1 = 2^{13} + 2^{12} + 2^5 + 2^4 + 2^3 + 2^0 = 11000000111001 .$$

Binary notation uses two digits $0, 1$ for the coefficients and the powers of two are indicated by the places of the coefficients.

Another popular number system is the *hexadecimal* or base 16 number system which is used in computing. For this number system we need 16 "digits" and the convention is to use $0, 1, 2, \ldots, 8, 9, A, B, C, D, E, F$.

The base 60 number system is not as unfamiliar as it may seem, because we represent time, geographical coordinates and angles in that system. We are comfortable carrying out computations such as (1 hour 40 minutes) plus (1 hour 30 minutes) equals (3 hours 10 minutes).

**Table 8.2** Babylonian triples in base 60

| a | c |
|---|---|
| ⟨1⟩⟨59⟩ | ⟨2⟩⟨49⟩ |
| ⟨1⟩⟨16⟩⟨41⟩ | ⟨1⟩⟨50⟩⟨49⟩ |
| ⟨3⟩⟨31⟩⟨49⟩ | ⟨5⟩⟨09⟩⟨01⟩ |
| ⟨1⟩⟨05⟩ | ⟨1⟩⟨37⟩ |

Table 8.2 shows the values of $a, c$ that appear in the tablet in base 60 notation where $\langle d \rangle$ represents the $d$'th "digit" for $0 \leq d < 60$. The reader can check that these values are the same as the decimal values given in Table 8.1, for example:

$$(3 \times 60^2) + (31 \times 60^1) + (49 \times 60^0) = 12709$$
$$(5 \times 60^2) + (9 \times 60^1) + (1 \times 60^0) = 18541$$

The Babylonians did not have 60 distinct symbols for the digits. Instead, they used a hybrid system where the coefficients were represented with two symbols: one for the tens coefficient and the other for the ones coefficient, and the places of the coefficients indicated the powers of 60. Using ♡ for the tens coefficient and ◊ for the ones coefficient, the decimal number $(38 \times 60) + (16 \times 60^0) = 2296$ would be represented as:

$$\begin{matrix} & \diamond\diamond\diamond\diamond & & \diamond\diamond & \\ ♡♡♡ & \diamond\diamond\diamond\diamond & ♡ & \diamond\diamond\diamond\diamond & . \end{matrix}$$

## What Is the Surprise?

Frank P. Ramsey's theorem appeared to be a minor result in combinatorics. Surprisingly, the theorem was the foundation of an entirely new and challenging field of mathematics with many open problems. The nature of Ramsey theory is also surprising: if a set is large enough there exist regularities in its subsets.

I was introduced to Ramsey theory by the article by Marijn J. H. Heule and Oliver Kullmann on Pythagorean triples whose proof bears some similarity to the proof of the four-color theorem: the use of massive computing resources that is only successful after theoretical advances. Hence the title of their article: *The Science of Brute Force*.

Problems in combinatorics ask for specific numerical values, for example, $R(n)$ must be a specific positive integer. It is surprising that probabilistic methods have proved so fruitful in obtaining results in this field.

We tend to think that humans are smarter today then they used to be thousands of years ago. It can be a surprise to find out that four thousand years ago Babylonian mathematics was sufficiently advanced to discover that $\{12709, 13500, 18541\}$ is a Pythagorean triple.

## Sources

For an overview of Ramsey theory see [9], while an advanced presentation can be found in [20]. The section on the probabilistic method is based on [43, Example 4o] and [9, Chapter 4]. A database of Ramsey numbers can be found in [34].

The method of proof of the theorem on Pythagorean triples is explained in detail in [23]. See [4] for an introduction to logic and to SAT solving. The archive of my SAT solver for education [5] contains formulas for Schur triples, Ramsey graphs and van der Waerden's problem.

Section 8.7 is based upon [60], [42]. The sexagesimal number system is described in [63].

# Chapter 9
# Langford's Problem

C. Dudley Langford noticed that his son had arranged colored blocks as shown in Fig. 9.1. There is one block between the red blocks, two blocks between the blue blocks and three blocks between the green blocks.

**Fig. 9.1** Layout of blocks for Langford's problem

**Definition 9.1 (Langford's Problem $L(n)$)** Given the multiset[1] of positive integers:

$$\{1, 1, 2, 2, 3, 3, \ldots, n, n\},$$

can they be arranged in a sequence such that for $1 \leq i \leq n$ there are $i$ numbers between the two occurrences of $i$?

Figure 9.1 shows that 312132 is a solution for $L(3)$.

Section 9.1 restates Langford's problem using a mathematical formalism that facilitates solving the problem. Section 9.2 characterizes values of $n$ for which $L(n)$ is solvable and presents two proofs of the theorem. The first proof which is relatively simple uses the technique of double-counting: counting the same value in two different ways and equating the resulting formulas. The second proof is a clever induction but the "bookkeeping" involved requires careful attention to the details. Section 9.3 works out the solution for $L(4)$.

---

[1] A *multiset* or *bag* is like a set except that there may be more than one occurrence of an element.

© The Author(s) 2022
M. Ben-Ari, *Mathematical Surprises*, https://doi.org/10.1007/978-3-031-13566-8_9

## 9.1 Langford's Problem as a Covering Problem

Langford's problem can be posed using an array. For $L(3)$ there are six columns, one for each position at which the six numbers can be placed. There is one row for each possible placement of one of the numbers, that is, the two occurrences of $k$ must have $k$ numbers between them. There are four possible placements of 1's, three of 2's and two of 3's:

|   | 1 | 2 | 3 | 4 | 5 | 6 |
|---|---|---|---|---|---|---|
| 1 | 1 |   | 1 |   |   |   |
| 2 |   | 1 |   | 1 |   |   |
| 3 |   |   | 1 |   | 1 |   |
| 4 |   |   |   | 1 |   | 1 |
| 5 | 2 |   |   | 2 |   |   |
| 6 |   | 2 |   |   | 2 |   |
| 7 |   |   | 2 |   |   | 2 |
| 8 | 3 |   |   |   | 3 |   |
| 9 |   | 3 |   |   |   | 3 |

To solve the problem we need to select one row for the 1's in the sequence, one row for the 2's and one row for the 3's, such that if we stack these rows on top of each other, no column contains more than one number.

Row 9 needed not be considered because of symmetry: starting with row 9 just gives the reversal of the sequence obtained by starting with row 8.

Row 8 is the only one containing 3's so it must be chosen and the sequence is 3⌣⌣⌣3⌣. Any row with numbers in columns 1 and 5 can no longer be used, because only one number can be placed at each position. Let us denote the permissible and forbidden rows by:

$$1, 2, \cancel{3}, 4, \cancel{5}, \cancel{6}, 7, 8.$$

Row 7 is the only remaining row containing 2's so it must be chosen and the sequence is 3⌣2⌣32. Deleting rows that can no longer be used gives:

$$1, 2, \cancel{3}, \cancel{4}, \cancel{5}, \cancel{6}, 7, 8.$$

Choosing the only remaining row, row 2, gives the solution 312132:

|   | 1 | 2 | 3 | 4 | 5 | 6 |
|---|---|---|---|---|---|---|
| 2 |   | 1 |   | 1 |   |   |
| 7 |   |   | 2 |   |   | 2 |
| 8 | 3 |   |   |   | 3 |   |

The analysis has shown that this is the only solution, except for the symmetrical solution obtained by starting with row 9.

## 9.2 For Which Values of $N$ Is Langford's Problem Solvable?

**Theorem 9.1** $L(n)$ *has a solution if and only if* $n = 4k$ *or* $n = 4k + 3$.

We prove the forward direction of the theorem. Proof 1 shows that if $L(n)$ has a solution then $n = 4k$ or $n = 4k + 3$. Proof 2 shows the contrapositive: if $n = 4k + 1$ or $n = 4k + 2$ then $L(n)$ has no solution.

**Proof (1)** If the first occurrence of the number $k$ is at position $i_k$, the second occurrence is at position $i_k + k + 1$. For example, in 312132, the solution for $L(3)$, choosing $k = 2$ gives $i_k = 3$ and $i_k + k + 1 = 3 + 2 + 1 = 6$.

$S_n$, the sum of the positions of all the numbers, is:

$$
S_n = \sum_{k=1}^{n} i_k + \sum_{k=1}^{n} (i_k + k + 1)
$$

$$
= 2 \sum_{k=1}^{n} i_k + \sum_{k=1}^{n} (k + 1)
$$

$$
= 2 \sum_{k=1}^{n} i_k + \frac{n(n + 3)}{2}.
$$

But $S_n$ is simply $1 + 2 + 3 + \cdots + 2n$, so:

$$
S_n = \sum_{k=1}^{2n} k = \frac{2n(2n + 1)}{2}.
$$

Equating the two formulas for $S_n$ gives:

$$
2 \sum_{k=1}^{n} i_k + \frac{n(n + 3)}{2} = \frac{2n(2n + 1)}{2}
$$

$$
\sum_{k=1}^{n} i_k = \frac{1}{2} \left( \frac{2n(2n + 1)}{2} - \frac{n(n + 3)}{2} \right)
$$

$$
= \frac{3n^2 - n}{4}.
$$

The left-hand side is an integer since it is the sum of integers (the positions), so the right-hand side must also be an integer. When is $3n^3 - n$ divisible by 4? Factoring $3n^2 - n$ gives $n(3n - 1)$.

If $n$ is a multiple of 4, the product is divisible by 4.

When is $3n - 1$ divisible by 4? Any integer $n$ can be expressed as $n = 4i + j$ for $j = 0, 1, 2, 3$. If $3n - 1$ is divisible by 4, then so is $3(4i + j) - 1 = 12i + 3j - 1$. $12i$ is divisible by 4. For $j = \{0, 1, 2, 3\}$, $3j - 1 = \{-1, 2, 5, 8\}$ is divisible by 4 if and only if $j = 3$, that is, $n = 4i + 3$. □

To introduce the idea of the second proof consider what a solution for $n = 4$ might look like. In the following tables the positions of the occurrences of 4 are 1 and 6, and the positions of the occurrences of 2 are 5 and 8. In both cases, one position is odd and the other is even.

| 1 | 2 | 3 | 4 | 5 | 6 | 7 | 8 |
|---|---|---|---|---|---|---|---|
| 4 | 1 | 3 | 1 | 2 | 4 | 3 | 2 |
| * |   |   |   |   | * |   |   |

| 1 | 2 | 3 | 4 | 5 | 6 | 7 | 8 |
|---|---|---|---|---|---|---|---|
| 4 | 1 | 3 | 1 | 2 | 4 | 3 | 2 |
|   |   |   |   | * |   |   | * |

Let $k = 2m$ be an *even* number. If $i$ is the position of the first occurrence of $k$, then the position of the second occurrence is $i + k + 1$. The sum of the positions is:

$$i + (i + k + 1) = 2i + 2m + 1 = 2(i + m) + 1,$$

which is an odd number. For the sum of two numbers to be odd, one must be odd and the other even.

Let us now check the positions of the occurrences of odd numbers. The positions of the occurrences of 1 are 2 and 4, both even numbers, and the positions of the occurrences of 3 are 3 and 7, both odd numbers.

| 1 | 2 | 3 | 4 | 5 | 6 | 7 | 8 |
|---|---|---|---|---|---|---|---|
| 4 | 1 | 3 | 1 | 2 | 4 | 3 | 2 |
|   | * |   | * |   |   |   |   |

| 1 | 2 | 3 | 4 | 5 | 6 | 7 | 8 |
|---|---|---|---|---|---|---|---|
| 4 | 1 | 3 | 1 | 2 | 4 | 3 | 2 |
|   |   | * |   |   |   | * |   |

Let $k = 2m + 1$ be an *odd* number. The sum of the positions is:

$$i + (i + k + 1) = 2i + 2m + 1 + 1 = 2(i + m + 1),$$

which is an even number. For the sum of two numbers to be even, both must be odd or both even.

The positions $1, 2, \ldots, 2n - 1, 2n$ contain an equal number of even and odd positions. The two occurrences of a number in a row "cover" two positions. When the set of rows covers all the positions, they must cover an equal number of even positions and odd positions. Define the *parity* of a set of rows to be the difference between the number of even and odd positions covered. Initially, the parity is zero, and if the problem has a solution, the set of rows in the solution also has zero parity.

When two occurrences of an even number are placed, they cover one even position and one odd position, so the parity remains the same:

| 1 | 2 | 3 | 4 | 5 | 6 | 7 | 8 |
|---|---|---|---|---|---|---|---|
| 4 | 1 | 3 | 1 | 2 | 4 | 3 | 2 |
| −1 |   |   |   |   | +1 |   |   |

| 1 | 2 | 3 | 4 | 5 | 6 | 7 | 8 |
|---|---|---|---|---|---|---|---|
| 4 | 1 | 3 | 1 | 2 | 4 | 3 | 2 |
|   |   |   |   | −1 |   |   | +1 |

When two occurrences of an odd number are placed, the parity becomes $+2$ or $-2$, so we must be able to associate this pair with a pair of occurrences of *another* odd number that are placed at positions that balance out the parity:

| 1 | 2 | 3 | 4 | 5 | 6 | 7 | 8 |
|---|---|---|---|---|---|---|---|
| 4 | 1 | 3 | 1 | 2 | 4 | 3 | 2 |
| | +1 | | | +1 | | | |

| 1 | 2 | 3 | 4 | 5 | 6 | 7 | 8 |
|---|---|---|---|---|---|---|---|
| 4 | 1 | 3 | 1 | 2 | 4 | 3 | 2 |
| | | | −1 | | | −1 | |

We have shown that there can be a solution to Langford's problem if and only if *there is an even number of odd numbers in* $\{1, \ldots, n\}$! The theorem claims that if this is true then either $n = 4k$ or $n = 4k - 1$, and if not then either $n = 4k - 2$ or $4k - 3$.

**Proof (2)** The proof is by induction. There are four base cases:

- $n = 4k - 3 = 1$. In $\{1\}$ there is an odd number of odd numbers and there is no solution.
- $n = 4k - 2 = 2$. In $\{1, 2\}$ there is an odd number of odd numbers and there is no solution.
- $n = 4k - 1 = 3$. In $\{1, 2, 3\}$ there is an even number of odd numbers and we have seen that there is a solution.
- $n = 4k - 0$. In $\{1, 2, 3, 4\}$ there is an even number of odd numbers and Sect. 9.3 gives a solution.

The inductive hypothesis is that the theorem is true for $\{1, \ldots, 4k - j\}$, $k \geq 1$, $0 \leq j \leq 3$, and we will prove that it is true for $n = 4(k + 1) - j$.

- Add $4k + 1 = 4(k + 1) - 3$ to $\{1, \ldots, 4k\}$. By the inductive hypothesis for $4k = 4k - 0$ there is an even number of odd numbers. $4(k + 1) - 3$ is odd so there is now an odd number of odd numbers and there is no solution.
- Add $4k + 2 = 4(k + 1) - 2$ to $\{1, \ldots, 4k + 1\}$. By the inductive hypothesis for $4k + 1 = 4(k + 1) - 3$ there is an odd number of odd numbers. $4(k + 1) - 2$ is even so there is still an odd number of odd numbers and there is no solution.
- Add $4k + 3 = 4(k + 1) - 1$ to $\{1, \ldots, 4k + 2\}$. By the inductive hypothesis for $4k + 2 = 4(k + 1) - 2$ there is an odd number of odd numbers. $4(k + 1) - 1$ is odd so there is an even number of odd numbers and a solution likely exists.
- Add $4k + 4 = 4(k + 1) - 0$ to $\{1, 2, \ldots, 4k + 3\}$. By the inductive hypothesis for $4k + 3 = 4(k + 1) - 1$ there is an even number of odd numbers. $4(k + 1) - 0$ is even so there is an even number of odd numbers and a solution likely exists. □

## 9.3 Solution for $L(4)$

Here is the array for $L(4)$. Try to find the solution yourself.

|    | 1 | 2 | 3 | 4 | 5 | 6 | 7 | 8 |
|----|---|---|---|---|---|---|---|---|
| 1  | 1 |   | 1 |   |   |   |   |   |
| 2  |   | 1 |   | 1 |   |   |   |   |
| 3  |   |   | 1 |   | 1 |   |   |   |
| 4  |   |   |   | 1 |   | 1 |   |   |
| 5  |   |   |   |   | 1 |   | 1 |   |
| 6  |   |   |   |   |   | 1 |   | 1 |
| 7  | 2 |   |   | 2 |   |   |   |   |
| 8  |   | 2 |   |   | 2 |   |   |   |
| 9  |   |   | 2 |   |   | 2 |   |   |
| 10 |   |   |   | 2 |   |   | 2 |   |
| 11 |   |   |   |   | 2 |   |   | 2 |
| 12 | 3 |   |   |   | 3 |   |   |   |
| 13 |   | 3 |   |   |   | 3 |   |   |
| 14 |   |   | 3 |   |   |   | 3 |   |
| 15 |   |   |   | 3 |   |   |   | 3 |
| 16 | 4 |   |   |   |   | 4 |   |   |
| 17 |   | 4 |   |   |   |   | 4 |   |
| 18 |   |   | 4 |   |   |   |   | 4 |

By symmetry row 18 may be eliminated.

Choose row 16 and the sequence is 4⌣⌣⌣⌣4 ⌣⌣. Any row with an element in position 1 or position 6 can no longer be part of the solution.

~~1~~, 2, 3, ~~4~~, 5, ~~6~~, ~~7~~, 8, ~~9~~, 10, 11, ~~12~~, ~~13~~, 14, 15, 16, ~~17~~

Choose row 14 and the sequence is 4⌣3⌣⌣43⌣.

~~1~~, 2, ~~3~~, ~~4~~, ~~5~~, ~~6~~, ~~7~~, 8, ~~9~~, ~~10~~, 11, ~~12~~, ~~13~~, 14, ~~15~~, 16, ~~17~~

Choose row 8. The sequence is 423⌣243⌣.

~~1~~, ~~2~~, ~~3~~, ~~4~~, ~~5~~, ~~6~~, ~~7~~, 8, ~~9~~, ~~10~~, ~~11~~, ~~12~~, ~~13~~, 14, ~~15~~, 16, ~~17~~

All of the choices for 1's have been eliminated so we must backtrack.

Instead of row 8 choose row 11 and the sequence is 4⌣3⌣2432.

~~1~~, 2, ~~3~~, ~~4~~, ~~5~~, ~~6~~, ~~7~~, ~~8~~, ~~9~~, ~~10~~, 11, ~~12~~, ~~13~~, 14, ~~15~~, 16, ~~17~~

Choose row 2 and we have a solution 41312432.

Continue backtracking to see if there is another solution.

Instead of row 14 choose row 15 and the sequence is 4⌣⌣3⌣4⌣3.

~~1~~, ~~2~~, 3, ~~4~~, 5, ~~6~~, ~~7~~, 8, ~~9~~, ~~10~~, ~~11~~, ~~12~~, ~~13~~, ~~14~~, 15, 16, ~~17~~

Row 8 must be chosen and the sequence is 42⌣324⌣3.

1̸, 2̸, 3̸, 4̸, 5̸, 6̸, 7̸, 8, 9̸, 1̸0, 1̸1, 1̸2, 1̸3, 1̸4, 15, 16, 1̸7

All of the choices for 1's have been eliminated so again we backtrack.

Instead of row 16 choose row 17 and the sequence is ⌣4⌣⌣⌣⌣4⌣.

1, 2̸, 3, 4, 5̸, 6, 7, 8̸, 9, 1̸0, 11, 12, 1̸3, 1̸4, 15, 1̸6, 17

Choose row 15 and the sequence is ⌣4⌣3⌣⌣43.

1, 2̸, 3, 4̸, 5̸, 6̸, 7̸, 8, 9, 1̸0, 1̸1, 1̸2, 1̸3, 1̸4, 15, 1̸6, 17

Row 9 must be chosen and the sequence is ⌣423⌣243.

1, 2̸, 3̸, 4̸, 5̸, 6̸, 7̸, 8̸, 9, 1̸0, 1̸1, 1̸2, 1̸3, 1̸4, 15, 1̸6, 17

All of the choices for 1's have been eliminated. We can backtrack one last time.

Instead of row 15 choose row 12 and the sequence is 34⌣⌣⌣3⌣4.

1̸, 2̸, 3̸, 4̸, 5̸, 6̸, 7̸, 8̸, 9, 1̸0, 1̸1, 12, 1̸3, 1̸4, 1̸5, 1̸6, 17

Again, all of the choices for 1's have been eliminated.

Therefore the only solution is 41312432.

## What Is the Surprise?

The source of the inspiration for a mathematical theorem can be surprising. Langford noticed a pattern in his son's colored blocks which led to the interesting Thm. 9.1. Students should also be introduced to the fact that a theorem can have many completely different proofs.

## Sources

This chapter is based on [35]. [12] shows how to find a solution for $n = 4k$ and $n = 4k + 3$.

# Chapter 10
# The Axioms of Origami

Origami, the art of paper folding, was developed several centuries ago in Japan and now has a worldwide following. In the late twentieth century the mathematical theory of origami was developed. Its foundation is a set of seven axioms, the *Huzita–Hatori axioms*, named after Humiaki Huzita who formalized the first six axioms and Koshiro Hatori who found the seventh. Jacques Justin published all seven axioms several years before Huzita and Hatori, and Margherita P. Beloch formulated the sixth axiom in 1936. Nevertheless, the axioms as known as the Huzita-Hatori axioms.

In a sequence of three chapters we will explore the mathematics of origami. This chapter presents the axioms, Chap. 11 connects origami with the roots of polynomials and Chap. 12 shows that constructions with origami can solve problems that are impossible using a straightedge and compass.

This chapter contains a section for each of the seven axioms. Following a statement of an axiom and a diagram of the *fold* it specifies, the equations of the fold and the points of intersection are developed using analytic geometry. A fold can also be defined as a *geometric locus*, the set of all points satisfying some property. The term fold comes from the origami operation of folding a piece of paper, but here it is used to refer the geometric line that would be created by folding the paper.

Folds result in *reflections*. Given a point $p$, its reflection around a fold $l$ results in a point $p'$ such that $l$ is the perpendicular bisector of the line segment $\overline{pp'}$ (Fig. 10.1).

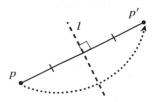

**Fig. 10.1** The fold is the perpendicular bisector of the line connecting a point and its reflection

© The Author(s) 2022
M. Ben-Ari, *Mathematical Surprises*, https://doi.org/10.1007/978-3-031-13566-8_10

## 10.1  Axiom 1

**Axiom 10.1** Given two distinct points $p_1 = (x_1, y_1)$, $p_2 = (x_2, y_2)$, there is a unique fold $l$ that passes through both of them (Fig. 10.2).

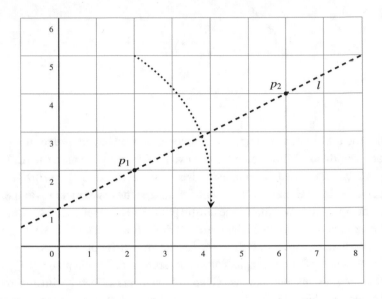

**Fig. 10.2** Axiom 1

**Derivation of the equation of the fold:** The equation of the fold $l$ is derived from the coordinates of $p_1$ and $p_2$. The slope is the quotient of the differences of the coordinates and the intercept is derived from $p_1$:

$$y - y_1 = \frac{y_2 - y_1}{x_2 - x_1}(x - x_1). \tag{10.1}$$

*Example 10.1* Let $p_1 = (2, 2)$, $p_2 = (6, 4)$. The equation of $l$ is:

$$y - 2 = \frac{4 - 2}{6 - 2}(x - 2)$$
$$y = \frac{1}{2}x + 1.$$

## 10.2 Axiom 2

**Axiom 10.2** Given two distinct points $p_1 = (x_1, y_1)$, $p_2 = (x_2, y_2)$, there is a unique fold $l$ that places $p_1$ onto $p_2$ (Fig. 10.3).

The fold is the geometric locus of all points equidistant from $p_1$ and $p_2$.

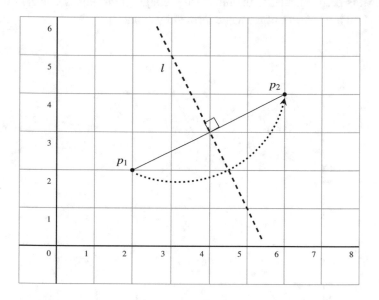

**Fig. 10.3** Axiom 2

**Derivation of the equation of the fold:** The fold $l$ is the perpendicular bisector of $\overline{p_1 p_2}$. Its slope is the negative reciprocal of the slope of the line connecting $p_1$ and $p_2$. $l$ passes through the midpoint between the points:

$$y - \frac{y_1 + y_2}{2} = -\frac{x_2 - x_1}{y_2 - y_1}\left(x - \frac{x_1 + x_2}{2}\right). \tag{10.2}$$

*Example 10.2* Let $p_1 = (2, 2)$, $p_2 = (6, 4)$. The equation of $l$ is:

$$y - \left(\frac{2+4}{2}\right) = -\frac{6-2}{4-2}\left(x - \left(\frac{2+6}{2}\right)\right)$$
$$y = -2x + 11 .$$

## 10.3 Axiom 3

**Axiom 10.3** Given two lines $l_1, l_2$, there is a fold $l$ that places $l_1$ onto $l_2$ (Fig. 10.4).

The fold is the geometric locus of the points that are equidistant from $l_1$ and $l_2$, where the distance from a point to a line is the length of the line segment through the point that is perpendicular to the line. Using congruent triangles it is easy to show that the fold is a bisector of the angle formed by $l_1$ and $l_2$.

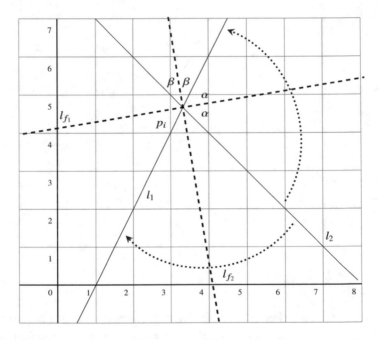

**Fig. 10.4** Axiom 3

**Derivation of the equation of the fold:**
$l_1, l_2$ *are parallel:* Let $l_1$ be $y = mx + b_1$ and let $l_2$ be $y = mx + b_2$. The fold is the line parallel to $l_1$ and $l_2$ that is halfway between them:

$$y = mx + \frac{b_1 + b_2}{2}.$$

$l_1, l_2$ *intersect:* Let $l_1$ be $y = m_1x + b_1$ and let $l_2$ be $y = m_2x + b_2$. $p_i = (x_i, y_i)$, the point of intersection of the two lines, is:

$$m_1 x_i + b_1 = m_2 x_i + b_2$$
$$x_i = \frac{b_2 - b_1}{m_1 - m_2}$$
$$y_i = m_1 x_i + b_1 .$$

*Example 10.3* Let $l_1$ be $y = 2x - 2$ and let $l_2$ be $y = -x + 8$. Then $p_i = (x_i, y_i)$ is:

$$x_i = \frac{8 - (-2)}{2 - (-1)} = \frac{10}{3} \approx 3.33$$
$$y_i = 2 \cdot \frac{10}{3} - 2 = \frac{14}{3} \approx 4.67 .$$

The fold is the bisector of the angle formed by $l_1$ and $l_2$ at their point of intersection. There are two possible folds since there are two pairs of vertical angles. We need to determine the slopes of the angle bisectors. If the angle of line $l_1$ relative to the x-axis is $\theta_1$ and the angle of line $l_2$ relative to the x-axis is $\theta_2$, then the fold is the line which makes an angle of $\theta_b = (\theta_1 + \theta_2)/2$ relative to the x-axis.

Let $m_1 = \tan \theta_1, m_2 = \tan \theta_2$. By Thm. A.9, $m_s$, the slope of the line making an angle of $\theta_1 + \theta_2$ relative to the x-axis, is:

$$m_s = \tan(\theta_1 + \theta_2) = \frac{\tan \theta_1 + \tan \theta_2}{1 - \tan \theta_1 \tan \theta_2} = \frac{m_1 + m_2}{1 - m_1 m_2} .$$

By Thm. A.10, $m_b$, the slope of the angle bisector, is:

$$m_b = \tan \frac{\theta_1 + \theta_2}{2} = \frac{-1 \pm \sqrt{1 + \tan^2(\theta_1 + \theta_2)}}{\tan(\theta_1 + \theta_2)} = \frac{-1 \pm \sqrt{1 + m_s^2}}{m_s} .$$

*Example 10.4* For $y = 2x - 2$ and $y = -x + 8$, the slope of the angle bisector is:

$$m_s = \frac{2 + (-1)}{1 - (2 \cdot -1)} = \frac{1}{3}$$
$$m_b = \frac{-1 \pm \sqrt{1 + (1/3)^2}}{1/3} = -3 \pm \sqrt{10} \approx -6.16, \; 0.162 .$$

Let us derive the equation of the fold $l_{f_1}$ with the positive slope. From Example 10.3, the coordinates of the intersection of the two lines are $(10/3, 14/3)$. Therefore:

$$\frac{14}{3} = (-3 + \sqrt{10}) \cdot \frac{10}{3} + b_i$$
$$b_i = \frac{44 - 10\sqrt{10}}{3}$$
$$y = (-3 + \sqrt{10})x + \frac{44 - 10\sqrt{10}}{3} \approx 0.162x + 4.13 .$$

## 10.4 Axiom 4

**Axiom 10.4** Given a point $p_1$ and a line $l_1$, there is a unique fold $l$ perpendicular to $l_1$ that passes through point $p_1$ (Fig. 10.5).

The fold is the geometric locus of all points on the line perpendicular to $l_1$ that passes through $p_1$.

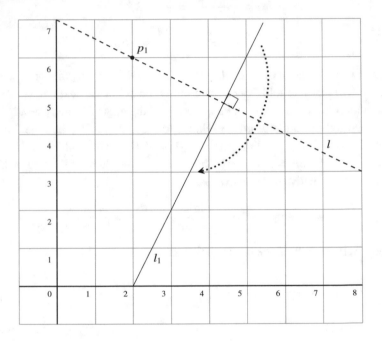

**Fig. 10.5** Axiom 4

**Derivation of the equation of the fold:** Let $l_1$ be $y = m_1 x + b_1$ and let $p_1 = (x_1, y_1)$. $l$ is perpendicular to $l_1$ so its slope is $-(1/m_1)$. Since it passes through $p_1$ we can compute the intercept $b$ and write down its equation:

$$y_1 = -\frac{1}{m}x_1 + b$$
$$b = \frac{(my_1 + x_1)}{m}$$
$$y = -\frac{1}{m}x + \frac{(my_1 + x_1)}{m}.$$

*Example 10.5* Let $p_1 = (2, 6)$ and let $l_1$ be $y = 2x - 4$. The equation of the fold $l$ is:

$$y = -\frac{1}{2}x + \frac{2 \cdot 6 + 2}{2} = -\frac{1}{2}x + 7.$$

## 10.5 Axiom 5

**Axiom 10.5** Given two points $p_1, p_2$ and a line $l_1$, there is a fold $l$ that places $p_1$ onto $l_1$ and passes through $p_2$ (Fig. 10.6).

Since the fold passes through $p_2$ and $p_2$ is on the perpendicular bisector of $\overline{p_1 p_1'}$, the geometric locus of the reflection of $p_1$ is the circle centered at $p_2$ with radius $\overline{p_1 p_2}$. The fold is constrained so that the reflection $p_1'$ is on the given line $l_1$.

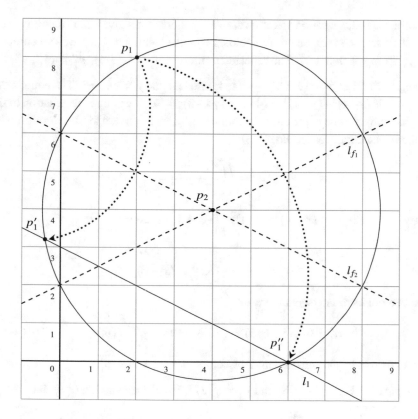

**Fig. 10.6** Axiom 5

**Derivation of the equations of the folds:** Let $l_1$ be $y = m_1 x + b_1$ and let $p_1 = (x_1, y_1)$, $p_2 = (x_2, y_2)$. The equation of the circle centered at $p_2$ with radius $\overline{p_1 p_2}$ is:

$$(x - x_2)^2 + (y - y_2)^2 = r^2, \quad \text{where}$$
$$r^2 = (x_2 - x_1)^2 + (y_2 - y_1)^2 .$$

Substituting the equation of $l_1$ into the equation for the circle gives:

$$(x - x_2)^2 + ((m_1x + b_1) - y_2)^2 = r^2$$
$$(x - x_2)^2 + (m_1x + (b_1 - y_2))^2 = r^2 ,$$

and we obtain a quadratic equation for the $x$-coordinates of the possible intersections:

$$x^2(1 + m_1^2) + 2(-x_2 + m_1(b - y_2))x + (x_2^2 + (b_1^2 - 2b_1y_2 + y_2^2) - r^2) = 0 .$$
$$(10.3)$$

Since a quadratic equation has at most two solutions, for a given pair of points and a line there may be zero, one or two folds. From the solutions $x_1', x_1''$ we can compute $y_1', y_1''$ from $y = m_1x + b_1$. The reflected points are $p_1' = (x_1', y_1')$, $p_1'' = (x_1'', y_1'')$.

*Example 10.6* Let $p_1 = (2, 8)$, $p_2 = (4, 4)$ and let $l_1$ be $y = -\frac{1}{2}x + 3$. The equation of the circle is $(x-4)^2+(y-4)^2 = (4-2)^2+(4-8)^2 = 20$. Substitute the equation of the line into the equation of the circle to obtain a quadratic equation for the $x$-coordinates of the intersections (or use Eq. 10.3):

$$(x - 4)^2 + \left(\left(-\frac{1}{2}x + 3\right) - 4\right)^2 = 20$$

$$(x - 4)^2 + (-1)^2 \cdot \left(\frac{1}{2}x + 1\right)^2 - 20 = 0$$

$$5x^2 - 28x - 12 = 0$$

$$(5x + 2)(x - 6) = 0 .$$

The two points of intersection are:

$$p_1' = (-2/5, 16/5) = (-0.4, 3.2), \quad p_1'' = (6, 0) .$$

The folds will be the perpendicular bisectors of $\overline{p_1p_1'}$ and $\overline{p_1p_1''}$.

*Example 10.7* For $p_1 = (2, 8)$ and $p_1' = (-2/5, 16/5)$ the equation of $l_{f_1}$ is:

$$y - \frac{8 + (16/5)}{2} = -\frac{(-2/5) - 2}{(16/5) - 8}\left(x - \frac{2 + (-2/5)}{2}\right)$$
$$y = -\frac{1}{2}x + 6 .$$

*Example 10.8* For $p_1 = (2, 8)$ and $p_1'' = (6, 0)$ the equation of $l_{f_2}$ is:

$$y - \frac{8 + 0}{2} = -\frac{6 - 2}{0 - 8}\left(x - \frac{2 + 6}{2}\right)$$
$$y = \frac{1}{2}x + 2 .$$

## 10.6 Axiom 6

**Axiom 10.6** Given two points $p_1, p_2$ and two lines $l_1, l_2$, there is a fold $l$ that places $p_1$ onto $l_1$ and places $p_2$ onto $l_2$ (Fig. 10.7).

A fold that places $p_i$ onto $l_i$ is a line $l_f$ such that the distance from $p_i$ to $l_f$ is equal to the distance from $l_i$ to $l_f$. The geometric locus of points equidistant from a point $p_i$ and a line $l_i$ is a *parabola*. $p_i$ is called the *focus* and $l_1$ is called the *directrix*. A fold is any line tangent to the parabola (Sect. 10.6.3).

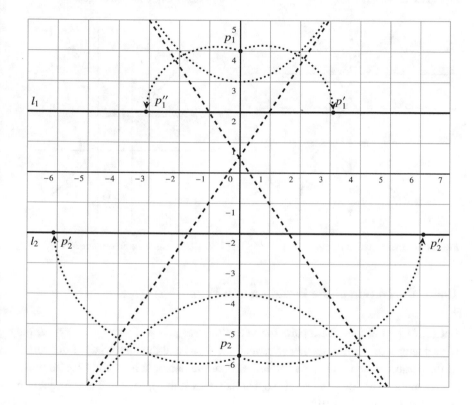

**Fig. 10.7** Axiom 6

For a fold to simultaneously place $p_1$ onto $l_1$ and $p_2$ onto $l_2$, it must be a tangent common to the two parabolas. There may be zero, one, two or three common tangents (Figs. 10.8a, 10.8b, 10.9a, 10.9b).

The formula for an arbitrary parabola is quite complex so we limit the presentation to parabolas whose axis of symmetry is the $x$- or $y$-axis.

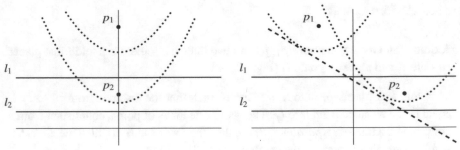

Fig. 10.8a  No common tangents                    Fig. 10.8b  One common tangent

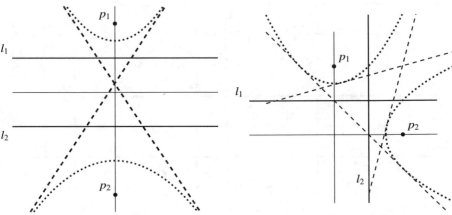

Fig. 10.9a  Two common tangents                   Fig. 10.9b  Three common tangents

## 10.6.1 Derivation of the Equation of a Fold

Let $(0, f)$ be the focus of a parabola with directrix $y = d$. Define $p = f - d$, the signed length of the line segment between the focus and the directrix.[1] If the vertex of the parabola is on the $x$-axis the equation of the parabola is $y = x^2/2p$. To move the parabola up or down the $y$-axis so that its vertex is at $(0, h)$, add $h$ to the equation of the parabola (Fig. 10.10):

$$y = \frac{x^2}{2p} + h.$$

---

[1] We have been using the notation $p_i$ for points; the use of $p$ here might be confusing but it is the standard notation. The formal name for $p$ is one-half the *latus rectum*.

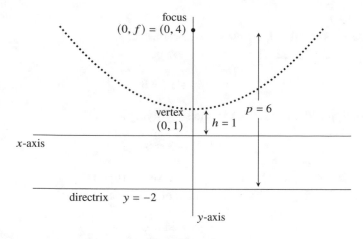

**Fig. 10.10** The elements in the definition of a parabola

Define $a = 2ph$ so that the equation of the parabola is:

$$y = \frac{x^2}{2p} + \frac{a}{2p}$$ (10.4a)

$$x^2 - 2py + a = 0.$$ (10.4b)

The equation of the parabola in Fig. 10.10 is $x^2 - 12y + 12 = 0$.

Substitute the equation of an *arbitrary* line $y = mx + b$ into Eq. 10.4b to obtain an equation for the points of intersection of the line and the parabola:

$$x^2 - 2p(mx + b) + a = 0$$
$$x^2 + (-2mp)x + (-2pb + a) = 0.$$

The line will be tangent to the parabola if and only if this quadratic equation has *exactly one solution* if and only if its discriminant is zero:

$$(-2mp)^2 - 4 \cdot 1 \cdot (-2pb + a) = 0$$ (10.5a)

$$m^2 p^2 + 2pb - a = 0.$$ (10.5b)

This is an equation with variables $m, b$ for the tangents to the parabola. To obtain the common tangents to both parabolas we must simultaneously solve the equations for the two parabolas.

*Example 10.9*

**Parabola 1:** Focus $(0, 4)$, directrix $y = 2$, vertex $(0, 3)$.
$p = 2$, $a = 2 \cdot 2 \cdot 3 = 12$. The equation of the parabola is:

$$x^2 - 4y + 12 = 0.$$

Substituting $p$ and $a$ into Eq. 10.5b and simplifying gives:

$$m^2 + b - 3 = 0.$$

**Parabola 2:** Focus $(0, -4)$, directrix $y = -2$, vertex $(0, -3)$.
$p = -2$, $a = 2 \cdot -2 \cdot -3 = 12$. The equation of the parabola is:

$$x^2 + 4y + 12 = 0.$$

Substituting $p$ and $a$ into Eq. 10.5b and simplifying gives:

$$m^2 - b - 3 = 0.$$

The solutions of the two equations:

$$m^2 + b - 3 = 0$$
$$m^2 - b - 3 = 0$$

are $m = \pm\sqrt{3} \approx \pm 1.73$ and $b = 0$. There are two common tangents:

$$y = \sqrt{3}x, \quad y = -\sqrt{3}x.$$

*Example 10.10*

**Parabola 1:** Unchanged.

**Parabola 2:** Focus $(0, -6)$, directrix $y = -2$, vertex $(0, -4)$.
$p = -4$, $a = 2 \cdot -4 \cdot -4 = 32$. The equation of the parabola is:

$$x^2 + 8y + 32 = 0.$$

Substituting $p$ and $a$ into Eq. 10.5b and simplifying gives:

$$2m^2 - b - 4 = 0.$$

The solutions of the two equations:

$$m^2 + b - 3 = 0$$
$$2m^2 - b - 4 = 0$$

are $m = \pm\sqrt{\dfrac{7}{3}} \approx \pm 1.53$ and $b = \dfrac{2}{3}$. There are two common tangents:

$$y = \sqrt{\frac{7}{3}}x + \frac{2}{3}, \quad y = -\sqrt{\frac{7}{3}}x + \frac{2}{3}.$$

*Example 10.11*

Let us now define a parabola whose axis of symmetry is the $x$-axis.

**Parabola 1:** Unchanged.

**Parabola 2:** Focus $(4, 0)$, directrix $x = 2$, vertex $(3, 0)$.

$p = 2$, $a = 2 \cdot 2 \cdot 3 = 12$. The equation of the parabola is:

$$y^2 - 4x + 12 = 0. \tag{10.6}$$

This is an equation with $x$ and $y^2$ instead of $x^2$ and $y$, so Eq. 10.5b can't be used and we must perform the derivation again.

Substitute the equation for a line into Eq. 10.6:

$$(mx + b)^2 - 4x + 12 = 0$$
$$m^2x^2 + (2mb - 4)x + (b^2 + 12) = 0.$$

Set the discriminant equal to zero and simplify:

$$(2mb - 4)^2 - 4m^2(b^2 + 12) = 0$$
$$-3m^2 - mb + 1 = 0.$$

If we try to solve the two equations:

$$m^2 + b - 3 = 0$$
$$-3m^2 - mb + 1 = 0,$$

we obtain a *cubic* equation with variable $m$:

$$m^3 - 3m^2 - 3m + 1 = 0. \tag{10.7}$$

Since a cubic equation has at least one and at most three real solutions, there can be one, two or three common tangents.

The formula for solving general cubic equations is quite complicated, so I used a calculator on the internet and obtained the three solutions:

$$m = 3.73, \quad m = -1, \quad m = 0.27.$$

From the form of Eq. 10.7 we might guess that $m = 1$ or $m = -1$ is a solution:

$$1^3 - 3 \cdot 1^2 - 3 \cdot 1 + 1 = -4$$
$$(-1)^3 - 3 \cdot (-1)^2 - 3 \cdot (-1) + 1 = 0.$$

Divide Eq. 10.7 by $m - (-1) = m + 1$ to obtain the quadratic equation $m^2 - 4m + 1$ whose roots are the other two solutions of the cubic equation $m = 2 \pm \sqrt{3} \approx 3.73, 0.27$.

## 10.6.2  Derivation of the Equations of the Reflections

We derive the position of the reflection $p_1' = (x_1', y_1')$ of $p_1 = (x_1, y_1)$ around a tangent line $l_t$ whose equation is $y = m_t x + b_t$. First, find the line $l_p$ with equation $y = m_p x + b_p$ that is perpendicular to $l_t$ and passes through $p_1$:

$$y = -\frac{1}{m_t}x + b_p$$

$$y_1 = -\frac{1}{m_t}x_1 + b_p$$

$$y = \frac{-x}{m_t} + \left(y_1 + \frac{x_1}{m_t}\right).$$

Next find the intersection $p_t = (x_t, y_t)$ of $l_t$ and $l_p$:

$$m_t x_t + b_t = \frac{-x_t}{m_t} + \left(y_1 + \frac{x_1}{m_t}\right)$$

$$x_t = \frac{\left(y_1 + \dfrac{x_1}{m_t} - b_t\right)}{\left(m_t + \dfrac{1}{m_t}\right)}$$

$$y_t = m_t x_t + b_t.$$

$p_t$ is the midpoint between $p_1$ and $p_1'$:

$$x_t = \frac{x_1 + x_1'}{2}, \qquad x_1' = 2x_t - x_1,$$
$$y_t = \frac{y_1 + y_1'}{2}, \qquad y_1' = 2y_t - y_1.$$

*Example 10.12* Let $l_t$ be $y = \sqrt{3}x + 0$ and let $p_1 = (0, 4)$:

$$x_t = \frac{\left(4 + \dfrac{0}{\sqrt{3}} - 0\right)}{\left(\sqrt{3} + \dfrac{1}{\sqrt{3}}\right)} = \sqrt{3}$$

$$y_t = \sqrt{3}\sqrt{3} + 0 = 3$$

$$x_1' = 2x_t - x_1 = 2\sqrt{3} \approx 3.46$$

$$y_1' = 2y_t - y_1 = 2.$$

### 10.6.3 Tangents to a Parabola

We wish to prove that the folds of Axiom 6 are tangents to the parabolas. Figure 10.11 shows five points $p_i$, $i = 1, \ldots, 5$, each point $p_i$ at a distance $a_i$ from both the focus and the directrix. Drop perpendicular lines from $p_i$ to the directrix and denote the intersections of these lines with the directrix by $p_i'$. By Axiom 2 there are folds $l_i$ through $p_i$ that place $p$ onto the directrix. The points $p_i'$ are the reflections of $p$ around the folds. The figure shows the fold $l_1$ through $p_1$ and the reflection $p_1'$.

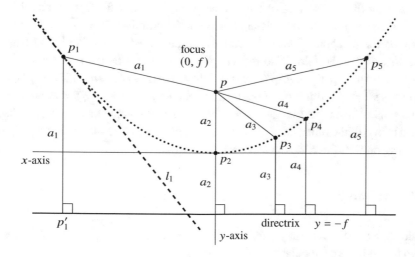

**Fig. 10.11** The tangent as a geometric locus

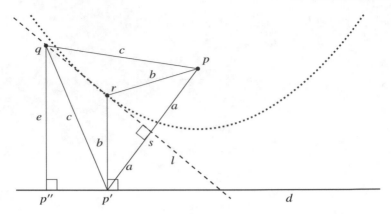

**Fig. 10.12** The proof that the fold is a tangent

**Theorem 10.7** *The folds of Axiom 6 are the tangents to the parabolas that are the loci of the points equidistant to the points $p_1, p_2$ and $l_1, l_2$, respectively.*

**Proof** In Fig. 10.12, the focus is $p$ and the directrix is $d$. $p'$ is a point on the directrix and $l$ is the fold that reflects $p$ onto $p'$. Let $s$ be the intersection of $\overline{pp'}$ and $l$. Then $\overline{ps} = \overline{p's} = a$ and $l \perp \overline{pp'}$ since $l$ is the perpendicular bisector of $\overline{pp'}$.

Let $r$ be the intersection of the line perpendicular to $d$ through $p'$ and the fold $l$. Then $\triangle psr \cong \triangle p'sr$ by side-angle-side. It follows that $\overline{pr} = \overline{p'r} = b$ so $r$ is a point on the parabola. Choose a point $p''$ on the directrix that is distinct from $p'$ and assume that the fold $l$ also reflects $p$ onto $p''$. Let $q$ be the intersection of the perpendicular to $d$ through $p''$ and the fold $l$. $\triangle psq \cong \triangle p'sq$ so $\overline{pq} = \overline{p'q} = c$. Denote $\overline{qp''} = e$. If $q$ is a point on the parabola then $e = \overline{qp''} = \overline{qp} = c$, but $c$ is the hypotenuse of the right triangle $\triangle qp''p'$ and it is not possible that the hypotenuse is equal to one of the other sides of the right triangle. Therefore the fold $l$ has only one intersection with the parabola and must be a tangent.                                            □

## 10.7  Axiom 7

**Axiom 10.8** Given a point $p_1$ and two lines $l_1$ and $l_2$, there is a fold $l$ that places $p_1$ onto $l_1$ and is perpendicular to $l_2$ (Fig. 10.13).

The fold is the geometric locus of all points on the line perpendicular to $l_2$ and equidistant from $p_1$ and $p'_1$, the reflection of $p_1$ onto $l_1$.

**Derivation of the equation of the fold:** Let $p_1 = (x_1, y_1)$, let $l_1$ be $y = m_1 x + b_1$ and let $l_2$ be $y = m_2 x + b_2$. Let $l_p$ be the line containing $\overline{p_1 p'_1}$. Since $l \perp l_2, l_p \perp l$, it follows that $l_p \parallel l_2$ and the equation of $l_p$ is $y = m_2 x + b_p$.

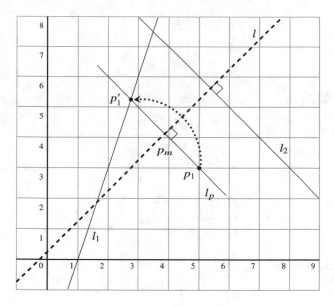

**Fig. 10.13** Axiom 7

$l_p$ passes through $p_1$ so $y_1 = m_2 x_1 + b_p$ and its equation is $y = m_2 x + (y_1 - m_2 x_1)$. The reflection $p'_1 = (x'_1, y'_1)$ is the intersection of $l_1$ and $l_p$:

$$m_1 x'_1 + b_1 = m_2 x'_1 + (y_1 - m_2 x_1)$$
$$x'_1 = \frac{y_1 - m_2 x_1 - b_1}{m_1 - m_2}$$
$$y'_1 = m_1 x'_1 + b_1 .$$

The equation of the midpoint $p_m = (x_m, y_m)$ of $l_p$ is:

$$(x_m, y_m) = \left( \frac{x_1 + x'_1}{2}, \frac{y_1 + y'_1}{2} \right) .$$

$l \perp l_2$ and it passes through $p_m$ so its equation is:

$$y = -\frac{1}{m_2} x + b_m,$$

where $b_m$ can be computed from $y = -\frac{1}{m_2} x + b_m$:

$$b_m = y_m + \frac{x_m}{m_2} .$$

The equation of the fold $l$ is therefore:

$$y = -\frac{1}{m_2}x + \left(y_m + \frac{x_m}{m_2}\right).$$

*Example 10.13* Let $p_1 = (5, 3)$, let $l_1$ be $y = 3x - 3$ and let $l_2$ be $y = -x + 11$. Then:

$$x_1' = \frac{3 - (-1) \cdot 5 - (-3)}{3 - (-1)} = \frac{11}{4}$$

$$y_1' = 3 \cdot \frac{11}{4} + (-3) = \frac{21}{4}$$

$$p_m = \left(\frac{5 + \frac{11}{4}}{2}, \frac{3 + \frac{21}{4}}{2}\right) = \left(\frac{31}{8}, \frac{33}{8}\right).$$

The equation of the fold $l$ is:

$$y = -\frac{1}{-1} \cdot x + \left(\frac{33}{8} + \frac{\frac{31}{8}}{-1}\right) = x + \frac{1}{4}.$$

## What Is the Surprise?

Origami, the art of paper folding, has been practiced for hundreds of years, so it is surprising that the mathematical formalization goes back only to the twentieth century. It is even more surprising that there is an axiomatization of paper folding. The mathematics of origami is an excellent way to learn analytic geometry, properties of parabolas and the concept of geometric locus.

## Sources

The axioms of origami are presented in [56]. Lang [26] gives descriptions of origami constructions. [31, Chap. 10] contains the detailed theory of the mathematics of origami, including the proof that two parabolas can have zero, one, two or three common tangents. The proof of Thm. 10.7 was shown to me by Oriah Ben-Lulu. I found that geometric software like Geogebra is useful for understanding the relation between the geometry and the algebra of the axioms.

A clear presentation of cubic equations can be found in [6, Chapters 1, 2].

# Chapter 11
# Lill's Method and the Beloch Fold

## 11.1 A Magic Trick

Construct a path consisting of four line segments $\{a_3 = 1, a_2 = 6, a_1 = 11, a_0 = 6\}$, starting from the origin along the positive direction of the $x$-axis and turning $90°$ counterclockwise between segments. Construct a second path as follows: construct a line from the origin at an angle of $63.4°$ and mark its intersection with $a_2$ by $P$. Turn left $90°$, construct a line and and mark its intersection with $a_1$ by $Q$. Turn left $90°$ once again, construct a line and note that it intersects the end of the first path at $(-10, 0)$ (Fig. 11.1).

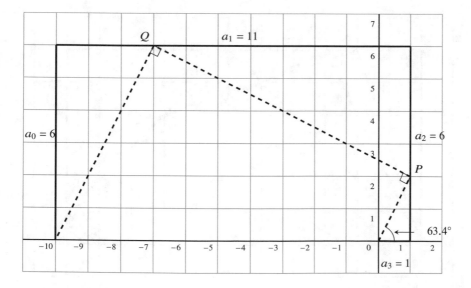

**Fig. 11.1** A magic trick

Compute the negation of the tangent of the angle at the start of the second path: $-\tan 63.4° = -2$. Substitute this value into the polynomial whose coefficients are the lengths of the segments of the first path:

$$p(x) = a_3x^3 + a_2x^2 + a_1x + a_0$$
$$= x^3 + 6x^2 + 11x + 6$$
$$p(-\tan 63.4°) = (-2)^3 + 6(-2)^2 + 11(-2) + 6 = 0.$$

We have found a root of the cubic polynomial $x^3 + 6x^2 + 11x + 6$!

Let us continue the example. The polynomial $p(x) = x^3 + 6x^2 + 11x + 6$ has three roots $-1, -2, -3$. Compute the arc tangent of the negation of the roots:

$$\alpha = -\tan^{-1}(-1) = 45°, \quad \beta = -\tan^{-1}(-2) \approx 63.4°, \quad \gamma = -\tan^{-1}(-3) \approx 71.6°.$$

For each angle the second path intersects the end of the first path (Fig. 11.2).

The value $-\tan 56.3 \approx -1.5$ is not a root of the equation. Fig. 11.3 shows the result of the application of the method for this angle. The second path does not intersect the line segment for the coefficient $a_0$ at $(-10, 0)$.

This example demonstrates a method discovered by Eduard Lill in 1867 for graphically finding the real roots of any polynomial. We are not actually finding the roots but verifying that a given value is a root.

Section 11.2 presents a formal specification of Lill's method (limited to cubic polynomials) and gives examples of how it works in special cases. A proof of the

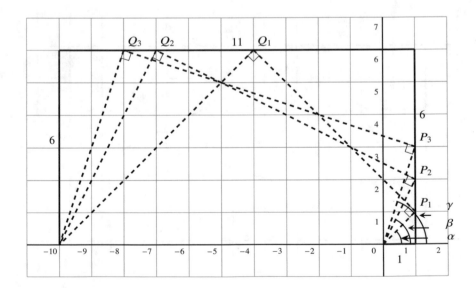

**Fig. 11.2** Lill's method for the three roots of the polynomial

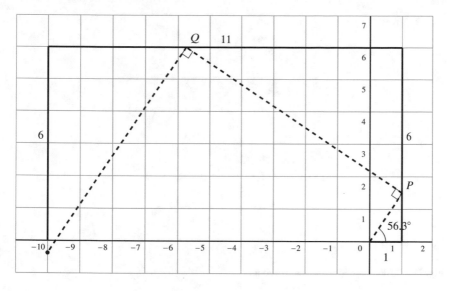

**Fig. 11.3** A path that does not lead to a root

correctness of Lill's method is given in Sect. 11.3. Section 11.4 shows how the method can be implemented using origami Axiom 6. This is called the Beloch fold and preceded the formalization of the axioms of origami by many years.

## 11.2 Specification of Lill's Method

### 11.2.1 Lill's Method as an Algorithm

- Start with an arbitrary cubic polynomial $p(x) = a_3x^3 + a_2x^2 + a_1x + a_0$.
- Construct the first path:

    – For each coefficient $a_3, a_2, a_1, a_0$ (in that order) construct a line segment of that length, starting at the origin $O = (0, 0)$ in the positive direction of the $x$-axis. Turn $90°$ counterclockwise between each segment.

- Construct the second path:

    – Construct a line from $O$ at an angle of $\theta$ with the positive $x$-axis that intersects $a_2$ at point $P$.
    – Turn $\pm 90°$ and construct a line from $P$ that intersects $a_1$ at $Q$.
    – Turn $\pm 90°$ and construct a line from $Q$ that intersects $a_0$ at $R$.
    – If $R$ is the end point of the first path then $-\tan \theta$ is a root of $p(x)$.

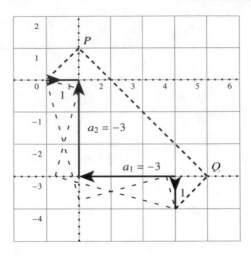

**Fig. 11.4** Lill's method with negative roots

- Special cases:

  - When constructing the line segments of the first path, if a coefficient is negative, construct the line segment *backwards*.
  - When constructing the line segments of the first path, if a coefficient is zero, do not construct a line segment but continue with the next ±90° turn.

- Notes:

  - The phrase *intersects* $a_i$ means *intersects the line segment* $a_i$ *or any extension of* $a_i$.
  - When building the second path choose to turn left or right by 90° so that there is an intersection with the next segment of the first path or its extension.

## 11.2.2 Negative Coefficients

Let us demonstrate Lill's method on the polynomial $p(x) = x^3 - 3x^2 - 3x + 1$ with negative coefficients (Sect. 10.6). Start by constructing a segment of length 1 to the right. Next, turn 90° to face up, but since the coefficient is negative, construct a segment of length 3 down, that is, in a direction opposite of the arrow. After turning 90° to the left, the coefficient is again negative, so construct a segment of length 3 to the right. Finally, turn downwards and construct a segment of length 1 (Fig. 11.4, the loosely dashed lines will be discussed in Sect. 11.2.4).

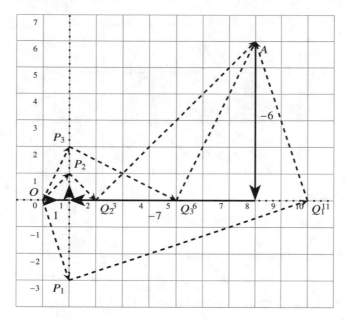

**Fig. 11.5** Lill's method with polynomials with zero coefficients

Start the second path with a line at $45°$ with the positive $x$-axis. It intersects the extension of the line segment for $a_2$ at $(1, 1)$. Turning $-90°$ (to the right), the line intersects the extension of the line segment for $a_1$ at $(5, -3)$. Turning $-90°$ again, the line intersects the end of the first path at $(4, -4)$. Since $-\tan 45° = -1$, we have found a root of the polynomial:

$$p(-1) = (-1)^3 - 3(-1)^2 - 3(-1) + 6 = 0.$$

### 11.2.3 Zero Coefficients

$a_2$, the coefficient of the $x^2$ term in the polynomial $x^3 - 7x - 6 = 0$, is zero. Construct a line segment of length 0, that is, do not construct a line, but still make the $\pm 90°$ turn as indicated by the arrow pointing up at $(1, 0)$ in Fig. 11.5. Turn again and construct a line segment of length $-7$, that is, of length 7 backwards, to $(8, 0)$. Finally, turn once more and construct a line segment of length $-6$ to $(8, 6)$.

The second paths with the following angles intersect the end of the first path:

$$-\tan^{-1}(-1) = 45°, \quad -\tan^{-1}(-2) \approx 63.4°, \quad -\tan^{-1} 3 \approx -71.6°.$$

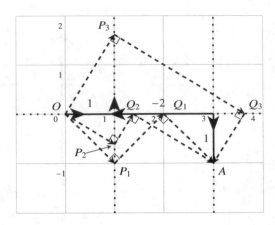

**Fig. 11.6** Lill's method with non-integer roots

We conclude that there are three real roots $\{-1, -2, 3\}$. Check:

$$(x + 1)(x + 2)(x - 3) = (x^2 + 3x + 2)(x - 3) = x^3 - 7x - 6.$$

### 11.2.4 Non-integer Roots

Figure 11.6 shows Lill's method for $p(x) = x^3 - 2x + 1$. The first path goes from $(0, 0)$ to $(1, 0)$ and then turns up. The coefficient of $x^2$ is zero so no line segment is constructed and the path turns left. The next line segment is of length $-2$ so it goes backwards from $(1, 0)$ to $(3, 0)$. Finally, the path turns down and a line segment of length 1 is constructed from $(3, 0)$ to $(3, -1)$.

It is easy to see that if the second path starts at an angle of $-45°$ it will intersect the first path at $(3, -1)$. Therefore, $-\tan^{-1}(-45)° = 1$ is a root. If we divide $p(x)$ by $x - 1$, we obtain the quadratic polynomial $x^2 + x - 1$ whose roots are:

$$\frac{-1 \pm \sqrt{5}}{2} \approx 0.62, \ -1.62.$$

There are two additional second paths: one starting at $-\tan^{-1} 0.62 \approx -31.8°$, and the other starting at $-\tan^{-1}(-1.62) \approx 58.3°$.

The polynomial $p(x) = x^3 - 3x^2 - 3x + 1$ (Sect. 11.2.2) has roots $2 \pm \sqrt{3} \approx 3.73, 0.27$. The corresponding angles are $-\tan^{-1} 3.73 \approx -75°$ and $-\tan^{-1} 0.27 \approx -15°$ as shown by the loosely dashed lines in Fig. 11.4.

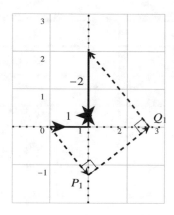

**Fig. 11.7** The cube root of two

### 11.2.5 The Cube Root of Two

To double a cube, compute $\sqrt[3]{2}$, a root of the cubic polynomial $x^3 - 2$. In the construction of the first path, turn left twice without constructing any line segments, because $a_2$ and $a_1$ are both zero. Then turn left again (to face down) and construct backwards (up) because $a_0 = -2$ is negative. The first segment of the second path is construct at an angle of $-\tan^{-1} \sqrt[3]{2} \approx -51.6°$ (Fig. 11.7).

## 11.3 Proof of Lill's Method

The proof is for monic cubic polynomials $p(x) = x^3 + a_2 x^2 + a_1 x + a_0$. If the polynomial is not monic, divide it by $a_3$ and the resulting polynomial has the same roots. In Fig. 11.8 the line segments of the first path are labeled with the coefficients and with $b_2, b_1, a_2 - b_2, a_1 - b_1$. In a right triangle if one acute angle is $\theta$ the other angle is $90° - \theta$. Therefore, the angle above $P$ and the angle to the left of $Q$ are equal to $\theta$. Here are the formulas for $\tan \theta$ as computed from the three triangles:

$$\tan \theta = \frac{b_2}{1} = b_2$$

$$\tan \theta = \frac{b_1}{a_2 - b_2} = \frac{b_1}{a_2 - \tan \theta}$$

$$\tan \theta = \frac{a_0}{a_1 - b_1} = \frac{a_0}{a_1 - \tan \theta (a_2 - \tan \theta)} \cdot$$

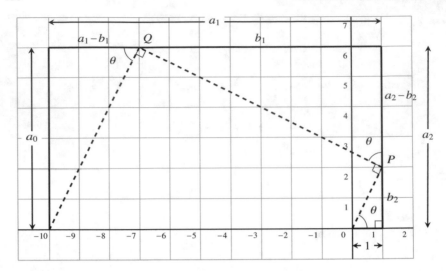

**Fig. 11.8** Proof of Lill's method

Simplify the last equation, multiply by $-1$ and absorb $-1$ into the powers:

$$(\tan\theta)^3 - a_2(\tan\theta)^2 + a_1(\tan\theta) - a_0 = 0$$
$$(-\tan\theta)^3 + a_2(-\tan\theta)^2 + a_1(-\tan\theta) + a_0 = 0.$$

It follows that $-\tan\theta$ is a real root of $p(x) = x^3 + a_2x^2 + a_1x + a_0$.

## 11.4 The Beloch Fold

Margharita P. Beloch discovered a remarkable connection between folding and Lill's method: one application of the operation later known as origami Axiom 6 generates a real root of a cubic polynomial. The operation is often called the *Beloch fold*.

Consider the polynomial $p(x) = x^3 + 6x^2 + 11x + 6$ (Sect. 11.1). Recall that a fold is the perpendicular bisector of the line segment between any point and its reflection around the fold. We want $\overline{RS}$ in Fig. 11.9 to be the perpendicular bisector of both $\overline{QQ'}$ and $\overline{PP'}$, where $Q', P'$ are the reflections of $Q, P$ around $\overline{RS}$, respectively.

Construct a line $a_2'$ parallel to $a_2$ at the same distance from $a_2$ as $a_2$ is from $P$, and construct a line $a_1'$ parallel to $a_1$ at the same distance from $a_1$ as $a_1$ is from $Q$. Apply Axiom 6 to simultaneously place $P$ at $P'$ on $a_2'$ and to place $Q$ at $Q'$ on $a_1'$. The fold $\overline{RS}$ is the perpendicular bisector of the lines $\overline{PP'}$ and $\overline{QQ'}$ so the angles at $R$ and $S$ are both right angles as required by Lill's method.

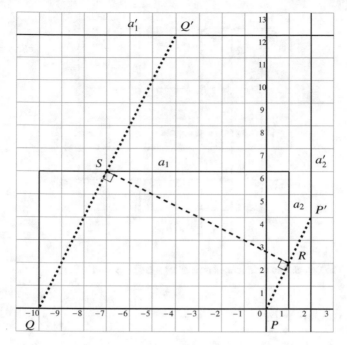

**Fig. 11.9** The Beloch fold for finding a root of $x^3 + 6x^2 + 11x + 6$

Figure 11.10 shows the Beloch fold for the polynomial $x^3 - 3x^2 - 3x + 1$ (Sect. 11.2.2). $a_2$ is the vertical line segment of length 3 whose equation is $x = 1$, and its parallel line is $a_2'$ whose equation is $x = 2$, because $P$ is at a distance of 1 from $a_2$. $a_1$ is the horizontal line segment of length 3 whose equation is $y = -3$, and its parallel line is $a_1'$ whose equation is $y = -2$ because $Q$ is at a distance of 1 from $a_1$. The fold $\overline{RS}$ is the perpendicular bisector of both $\overline{PP'}$ and $\overline{QQ'}$, and the line $\overline{PRSQ}$ is the same as the second path in Fig. 11.4.

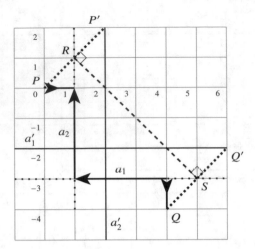

**Fig. 11.10** The Beloch fold for finding a root of $x^3 - 3x^2 - 3x + 1$

## What Is the Surprise?

Performing Lill's method as a magic trick never fails to surprise. It can be performed during a lecture using graphics software such as GeoGebra. It is also surprising that Lill's method, published in 1867, and Beloch's fold, published in 1936, preceded the axiomatization of origami by many years.

## Sources

This chapter is based on [8, 24, 40].

# Chapter 12
# Geometric Constructions Using Origami

This chapter shows that constructions with origami are more powerful than constructions with a straightedge and compass. We give two constructions for trisecting an angle, one by Hisashi Abe (Sect. 12.1) and the other by George E. Martin (Sect. 12.2), two constructions for doubling a cube, one by Peter Messer (Sect. 12.3) and the other by Marghareta P. Beloch (Sect. 12.4), and the construction of a nonagon, a regular polynomial with nine sides (Sect. 12.5).

## 12.1 Abe's Trisection of an Angle

**Construction:** Given an acute angle $\angle PQR$, construct $p$, the perpendicular to $\overline{QR}$ at $Q$. Construct $q$, a perpendicular to $p$ that intersects $\overline{PQ}$ at point $A$, and construct $r$, the perpendicular to $p$ at $B$ that is halfway between $Q$ and $A$. Using Axiom 6 construct the fold $l$ that places $A$ at $A'$ on $\overline{PQ}$ and $Q$ at $Q'$ on $r$. Let $B'$ be the reflection of $B$ around $l$. Construct lines through $\overline{QB'}$ and $\overline{QQ'}$ (Fig. 12.1).

**Theorem 12.1** $\angle PQB' = \angle B'QQ' = \angle Q'QR = \angle PQR/3$.

***Proof (1)*** $A', B', Q'$ are reflections around the line $l$ of the points $A, B, Q$ on the line $\overline{DQ}$, so they are on the reflected line $\overline{DQ'}$. By construction $\overline{AB} = \overline{BQ}$, $\angle ABQ' = \angle QBQ' = 90°$ and $\overline{BQ'}$ is a common side, so $\triangle ABQ' \cong \triangle QBQ'$ by side-angle-side. Therefore, $\angle AQQ' = \angle QAQ' = \alpha$ so $\triangle AQ'Q$ is isoceles (Fig. 12.2).

By reflection $\triangle AQ'Q \cong \triangle A'QQ'$, so $\triangle A'QQ'$ is also an isoceles triangle. $\overline{QB'}$, the reflection of $\overline{Q'B}$, is the perpendicular bisector of an isoceles triangle, so $\angle A'QB' = \angle Q'QB' = \angle QQ'B = \alpha$. By alternating interior angles, $\angle Q'QR = \angle QQ'B = \alpha$. Together we have:

$$\triangle PQB' = \angle A'QB' = \angle B'QQ' = \angle Q'QR = \alpha.$$ □

© The Author(s) 2022
M. Ben-Ari, *Mathematical Surprises*, https://doi.org/10.1007/978-3-031-13566-8_12

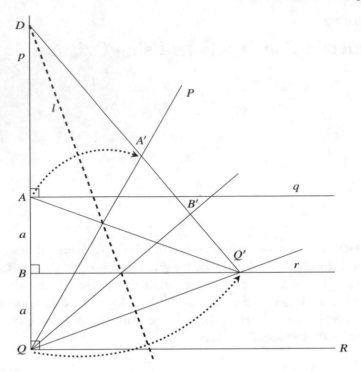

**Fig. 12.1** Abe's trisection of an angle

***Proof*** *(2)* Since $l$ is a fold it is the perpendicular bisector of $\overline{QQ'}$. Denote the intersection of $l$ with $\overline{QQ'}$ by $U$ and its intersection with $\overline{QB'}$ by $V$ (Fig. 12.2). $\triangle VUQ \cong \triangle VUQ'$ by side-angle-side since $\overline{VU}$ is a common side, the angles at $U$ are right angles and $\overline{QU} = \overline{Q'U}$. Therefore, $\angle VQU = \angle VQ'U = \alpha$ and $\angle Q'QR = \angle VQ'U = \alpha$ by alternating interior angles.

As in the first proof $A', B', Q'$ are all reflections around $l$, so they are on the line $\overline{DQ'}$ and $\overline{A'B'} = \overline{AB} = \overline{BQ} = \overline{B'Q'} = a$. Then $\triangle A'B'Q \cong \triangle Q'B'Q$ by side-angle-side and $\angle A'QB' = \angle Q'QB' = \alpha$.                                     □

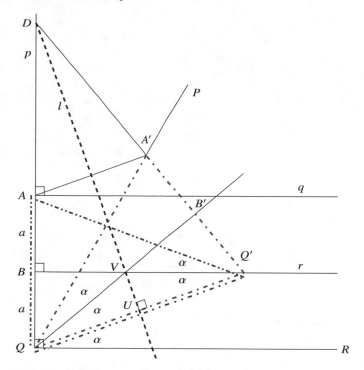

**Fig. 12.2** Proofs of Abe's trisection ($U, V$ are used in Proof 2)

## 12.2  Martin's Trisection of an Angle

**Construction:** Given an acute angle $\angle PQR$, let $M$ be the midpoint of $\overline{PQ}$. Construct $p$ the perpendicular to $\overline{QR}$ through $M$ and construct $q$ perpendicular to $p$ through $M$ so $q \parallel \overline{QR}$. Using Axiom 6 construct the fold $l$ that places $P$ at $P'$ on $p$ and $Q$ at $Q'$ on $q$. If more than one fold is possible choose the one that intersects $\overline{PM}$. Construct $\overline{PP'}$ and $\overline{QQ'}$ (Fig. 12.3).

**Theorem 12.2** $\angle Q'QR = \angle PQR/3$.

*Proof*  Denote the intersection of $\overline{QQ'}$ with $p$ by $U$ and its intersection with $l$ by $V$. Denote the intersection of $\overline{PQ}$ and $\overline{P'Q'}$ with $l$ by $W$. It is not immediate that $\overline{PQ}$ and $\overline{P'Q'}$ intersect $l$ at the same point. But $\triangle PWP' \sim \triangle QWQ'$ so the altitudes bisect both vertical angles $\angle PWP'$, $\angle QWQ'$ and they must be on the same line.

$\triangle QMU \cong \triangle PMP'$ by angle-side-angle since $\angle P'PM = \angle UQM = \beta$ by alternate interior angles, $\overline{QM} = \overline{MP} = a$ because $M$ is the midpoint of $\overline{PQ}$ and $\angle QMU = \angle PMP' = \gamma$ are vertical angles. Therefore, $\overline{P'M} = \overline{MU} = b$.

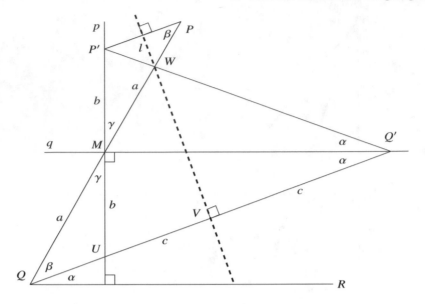

**Fig. 12.3** Martin's trisection of an angle

$\triangle P'MQ' \cong \triangle UMQ'$ by side-angle-side, since $\overline{P'M} = \overline{MU} = b$, the angles at $M$ are right angles and $\overline{MQ'}$ is a common side. Since the altitude of the isoceles triangle $\triangle P'Q'U$ is the bisector of $\angle P'Q'U$, it follows that $\angle P'Q'M = \angle UQ'M = \alpha$. Furthermore, $\angle UQ'M = \angle Q'QR = \alpha$ by alternate interior angles. $\triangle QWV \cong \triangle Q'WV$ by side-angle-side since $\overline{QV} = \overline{VQ'} = c$, the angles at $V$ are right angles and $\overline{VW}$ is a common side. Therefore:

$$\angle WQV = \beta = \angle WQ'V = 2\alpha$$
$$\angle PQR = \beta + \alpha = 3\alpha .$$  □

## 12.3 Messer's Doubling of a Cube

A cube of volume $V$ has sides of length $\sqrt[3]{V}$. A cube with twice the volume has sides of length $\sqrt[3]{2V} = \sqrt[3]{2}\sqrt[3]{V}$, so if we can construct $\sqrt[3]{2}$ we can multiply by the given length $\sqrt[3]{V}$ to double the cube.

**Construction:** Divide the side of a unit square into thirds as follows: Fold the square in half to locate the points $I = (0, 1/2)$ and $J = (1, 1/2)$. Next, construct the lines $\overline{AC}$ and $\overline{BJ}$ (Fig. 12.4). The point of intersection $K = (2/3, 1/3)$ can be obtained by solving the two equations $y = 1 - x$ and $y = x/2$.

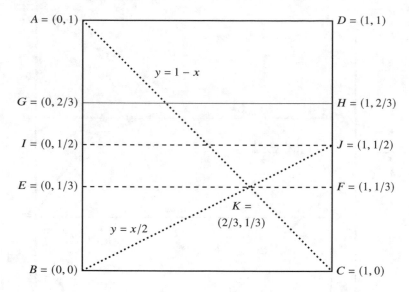

**Fig. 12.4** Dividing a length into thirds

   Construct $\overline{EF}$, the perpendicular to $\overline{AB}$ through $K$, and construct the reflection $\overline{GH}$ of $\overline{BC}$ around $\overline{EF}$. The side of the square has now been divided into thirds.

   Using Axiom 6 place $C$ at $C'$ on $\overline{AB}$ and $F$ at $F'$ on $\overline{GH}$. Denote by $L$ the point intersection of the fold with $\overline{BC}$ and denote by $b$ the length of $\overline{BL}$. Rename the length of the side of the square to $a + 1$ where $a = \overline{AC'}$. The length of $\overline{LC}$ is $(a + 1) - b$ (Fig. 12.5).

**Theorem 12.3** $\overline{AC'} = \sqrt[3]{2}$.

***Proof*** When the fold is performed the line segment $\overline{LC}$ is reflected onto the line segment $\overline{LC'}$ and $\overline{CF}$ is folded onto the line segment $\overline{C'F'}$. Therefore:

$$\overline{GC'} = a - \frac{a+1}{3} = \frac{2a-1}{3}. \tag{12.1}$$

Since $\angle FCL$ is a right angle, so is $\angle F'C'L$.

   $\triangle C'BL$ is a right triangle so by Pythagoras's Theorem:

$$1^2 + b^2 = ((a + 1) - b)^2 \tag{12.2a}$$

$$b = \frac{a^2 + 2a}{2(a + 1)}. \tag{12.2b}$$

   $\angle GC'F' + \angle F'C'L + \angle LC'B = 180°$ since they form the straight line $\overline{GB}$. Denote $\angle GC'F'$ by $\alpha$. Then:

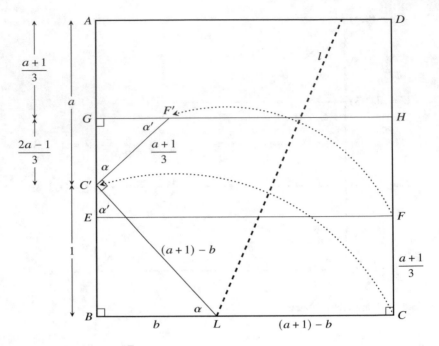

**Fig. 12.5** Construction of $\sqrt[3]{2}$

$$\angle LC'B = 180° - \angle F'C'L - \angle GC'F' = 180° - 90° - \alpha = 90° - \alpha,$$

which we denote by $\alpha'$. The triangles $\triangle C'BL$, $\triangle F'GC'$ are right triangles so $\angle C'LB = \alpha$ and $\angle C'F'G = \alpha'$. Therefore, $\triangle C'BL \sim \triangle F'GC'$ and:

$$\frac{\overline{BL}}{\overline{C'L}} = \frac{\overline{GC'}}{\overline{C'F'}}.$$

Using Eq. 12.1 we have:

$$\frac{b}{(a+1)-b} = \frac{\dfrac{2a-1}{3}}{\dfrac{a+1}{3}}.$$

Substituting for $b$ using Eq. 12.2b gives:

$$\frac{\dfrac{a^2+2a}{2(a+1)}}{(a+1)-\dfrac{a^2+2a}{2(a+1)}} = \frac{2a-1}{a+1}.$$

Simplify the equation to obtain $a^3 = 2$ and $a = \sqrt[3]{2}$.                    □

## 12.4 Beloch's Doubling of a Cube

Since the Beloch fold (Axiom 6) can solve cubic equations it is reasonable to conjecture that it can be used to double a cube. Here we give a direct construction that uses the fold.

**Construction:** Let $A = (-1, 0)$, $B = (0, -2)$. Let $p$ be the line $x = 1$ and let $q$ be the line $y = 2$. Use the Beloch fold to construct the fold $l$ that places $A$ at $A'$ on $p$ and $B$ at $B'$ on $q$. Denote the intersection of the fold and the $y$-axis by $Y$ and the intersection of the fold and the $x$-axis by $X$ (Fig. 12.6).

**Theorem 12.4** $\overline{OY} = \sqrt[3]{2}$.

***Proof*** The fold is the perpendicular bisector of both $\overline{AA'}$ and $\overline{BB'}$ so $\overline{AA'} \parallel \overline{BB'}$. By alternate interior angles $\angle YAO = \angle BXO = \alpha$. The labeling of the other angles in the figure follows from the properties of right triangles.

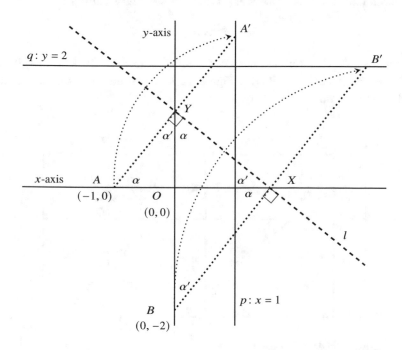

**Fig. 12.6** Beloch's doubling of the cube

$\triangle AOY \sim \triangle YOX \sim \triangle XOB$ and $\overline{OA} = 1$, $\overline{OB} = 2$ are given so:

$$\frac{\overline{OY}}{\overline{OA}} = \frac{\overline{OX}}{\overline{OY}} = \frac{\overline{OB}}{\overline{OX}}$$

$$\frac{\overline{OY}}{1} = \frac{\overline{OX}}{\overline{OY}} = \frac{2}{\overline{OX}}.$$

From the first and second ratios we have $\overline{OX} = \overline{OY}^2$ and from the first and third ratios we have $\overline{OY}\,\overline{OX} = 2$. Substituting for $\overline{OX}$ gives $\overline{OY}^3 = 2$ and $\overline{OY} = \sqrt[3]{2}$.  □

## 12.5 Construction of a Regular Nonagon

A nonagon (a regular polygon with nine sides) is constructed by deriving the cubic equation for its central angle and then solving the equation using Lill's method and the Beloch fold. The central angle is $\theta = 360°/9 = 40°$. By Thm. A.6:

$$\cos 3\theta = 4\cos^3 \theta - 3\cos \theta.$$

Let $x = \cos 40°$. Then for the nonagon the equation is $4x^3 - 3x + (1/2) = 0$ since $\cos 3 \cdot 40° = \cos 120° = -(1/2)$. Figure 12.7 shows the paths for the equation constructed according to Lill's method.

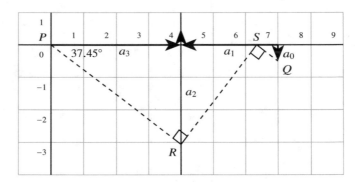

**Fig. 12.7** Lill's method for a nonagon

The second path starts from $P$ at an angle of approximately $-37.45°$. Turns of $90°$ at $R$ and then $-90°$ at $S$ cause the path to intersect the first path at its endpoint $Q$. Therefore, $x = -\tan(-37.45°) \approx 0.766$ is a root of $4x^3 - 3x + (1/2)$.

The root can be obtained using the Beloch fold. Construct the line $a_2'$ parallel to $a_2$ at the same distance from $a_2$ as $a_2$ is from $P$. Although the length of $a_2$ is zero, it still has a direction (upwards) so the parallel line can be constructed. Similarly, construct the line $a_1'$ parallel to $a_1$ at the same distance from $a_1$ as $a_1$ is from $Q$. The Beloch fold $\overline{RS}$ simultaneously places $P$ at $P'$ on $a_2'$ and $Q$ at $Q'$ on $a_1'$. This constructs the angle $\angle SPR = -37.45°$ (Fig. 12.8).

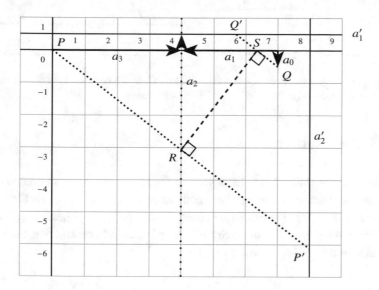

**Fig. 12.8** The Beloch fold for solving the equation of the nonagon

By Lill's method $-\tan(-37.45°) \approx 0.766$ and therefore $\cos\theta \approx 0.766$ is a root of the equation for the central angle $\theta$. We finish the construction of the nonagon by constructing $\cos^{-1} 0.766 \approx 40°$.

The right triangle $\triangle ABC$ with $\angle CAB \approx 37.45°$ and $\overline{AB} = 1$ has opposite side $\overline{BC} \approx 0.766$ by definition of tangent (Fig. 12.9a). Fold $\overline{CB}$ onto the $\overline{AB}$ so that the reflection of $C$ is $D$ and $\overline{DB} = 0.766$. Extend $\overline{DB}$ and construct $E$ so that $\overline{DE} = 1$. Fold $\overline{DE}$ to reflect $E$ at $F$ on the extension of $\overline{BC}$ (Fig. 12.9b). Then:

$$\angle BDF = \cos^{-1} \frac{0.766}{1} \approx 40°.$$

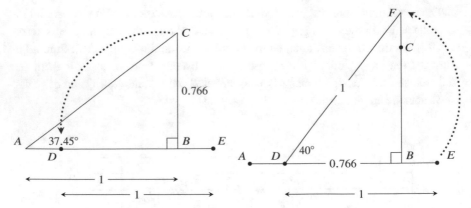

**Fig. 12.9a** The tangent that is the solution of the equation for the nonagon

**Fig. 12.9b** The cosine of the central angle of the nonagon

## What Is the Surprise?

We saw in Chaps. 2 and 3 that tools such as the neusis can perform constructions that cannot be done with a straightedge and compass. It is surprising that trisecting an angle and doubling a cube can be constructed using only paper folding. Roger C. Alperin has developed a hierarchy of four methods of construction each more powerful than the previous one.

## Sources

This chapter is based on [2, 26, 31, 36].

# Chapter 13
# A Compass Is Sufficient

In 1797 Lorenzo Mascheroni proved that any construction carried out with a straight-edge and compass can be carried out with only a compass. Later it came to light that this theorem had already been proved by Georg Mohr in 1672. After explaining in Sect. 13.1 what is meant by performing a construction with only a compass, the proof is presented in stages starting with four auxiliary constructions: reflection of a point (Sect. 13.2), construction of a circle with a given radius (Sect. 13.3), addition and subtraction of line segments (Sect. 13.4) and construction of a line segment as a ratio of segments (Sect. 13.5). Section 13.6 shows how to find the intersection of two lines and Sect. 13.7 shows how to find the intersection of a line and a circle.

## 13.1 What Is a Construction With Only a Compass?

Figure 13.1a shows the construction of an equilateral triangle using a straightedge and compass. How can we construct a triangle without the line segments $\overline{AB}, \overline{AC}, \overline{BC}$? A line segment is defined by two points, so it is sufficient to construct these points in order to obtain a construction equivalent to the one with a straightedge (Fig. 13.1b). There is no need to actually *see* the line segments. There will be lines in the figures in this chapter, but they are used only to understand the construction and the proof of its correctness. It is important to convince yourself that the construction itself uses only a compass.

A construction using a straightedge and compass is a sequence of three operations:

- Find the point of intersection of two lines.
- Find the point(s) of intersection of a line and a circle.
- Find the point(s) of intersection of two circles.

The third operation can be done with only a compass. We need to show that the first two operations can be done with a compass alone.

© The Author(s) 2022
M. Ben-Ari, *Mathematical Surprises*, https://doi.org/10.1007/978-3-031-13566-8_13

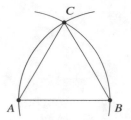

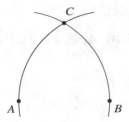

**Fig. 13.1a** Construction of an equilateral triangle with a straightedge and a compass

**Fig. 13.1b** Construction of an equilateral triangle with only a compass

Notation:

- $c(O, A)$: the circle with center $O$ through point $A$.
- $c(O, r)$: the circle with center $O$ and radius $r$.
- $c(O, \overline{AB})$: the circle with center $O$ and radius the length of line segment $\overline{AB}$.

## 13.2 Reflection of a Point

**Definition 13.1** A point $C'$ is a *reflection* of the point $C$ around a line segment $\overline{AB}$ if and only if $\overline{AB}$ (or the line containing $\overline{AB}$) is the perpendicular bisector of the line segment $\overline{CC'}$.

**Theorem 13.1** *Given a line $\overline{AB}$ and a point $C$ not on $\overline{AB}$, it is possible to build $C'$, the reflection of $C$ around $\overline{AB}$.*

**Proof** Construct a circle centered on $A$ passing through $C$ and a circle centered on $B$ passing through $C$. The other intersection of the two circles is the point $C'$ which is the reflection of $C$ (Fig. 13.2). $\triangle ABC \cong \triangle ABC'$ by side-side-side since $\overline{AC}, \overline{AC'}$ are radii of the same circle, as are $\overline{BC}, \overline{BC'}$ and $\overline{AB}$ is a common side. Therefore, $\angle CAB = \angle C'AB$ so $\overline{AB}$ is the angle bisector of $\angle CAC'$. But $\triangle CAC'$ is an isosceles triangle and the angle bisector $\overline{AB}$ is also the perpendicular bisector of $\overline{CC'}$, the base of $\triangle CAC'$. By definition $C'$ is the reflection of $C$ around $\overline{AB}$. □

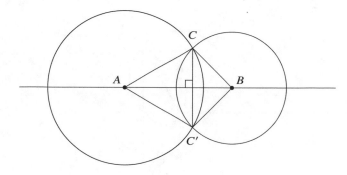

**Fig. 13.2** Construction of a reflection

## 13.3 Construction of a Circle With a Given Radius

**Theorem 13.2** *Given points $A, B, C$ it is possible to construct $c(A, \overline{BC})$, the circle centered at $A$ with radius $\overline{BC}$.*

**Proof** Construct $c(A, B)$ and $c(B, A)$ and let $X, Y$ be their points of intersection (Fig. 13.3). $A$ is the reflection of $B$ around $\overline{XY}$ since $\triangle YAX \cong \triangle YBX$ by side-side-side. By Thm. 13.1 construct $C'$, the reflection of $C$ around $\overline{XY}$ and then construct $c(A, \overline{AC'})$ (Fig. 13.4).

$\overline{XY}$ is the perpendicular bisector of $\overline{CC'}$ and $\overline{AB}$. Denote the intersection of $\overline{XY}$ and $\overline{AB}$ by $D$ and the intersection of $\overline{XY}$ and $\overline{CC'}$ by $E$. Then $\overline{C'E} = \overline{EC}$, $\overline{AD} = \overline{DB}$ and $\angle DEC = \angle DEC'$ is a right angle, so $\triangle DEC \cong \triangle DEC'$ by side-angle-side. Therefore, $\overline{DC} = \overline{DC'}$ and $\angle ADC' = \angle BDC$ (they are complementary to $\angle EDC' = \angle EDC$). It follows that $\triangle ADC' \cong \triangle BDC$ by side-angle-side so $\overline{AC'} = \overline{BC}$. □

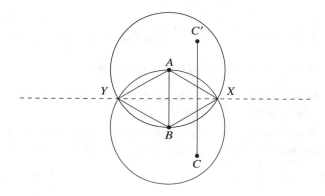

**Fig. 13.3** Construction of a circle with a given radius (1)

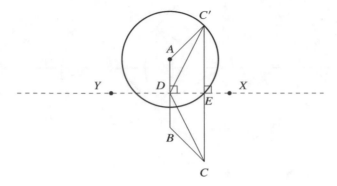

**Fig. 13.4** Construction of a circle with a given radius (2)

## 13.4  Addition and Subtraction of Line Segments

**Theorem 13.3** *Given a line segment $\overline{PQ}$ of length a and a line segment $\overline{RS}$ of length b, it is possible to construct line segments $\overline{QT}, \overline{QU}$ such that $\overline{PTQU}$ is a line segment, the length of $\overline{PT}$ is a − b and the length of $\overline{PU}$ is a + b (Fig. 13.5).*

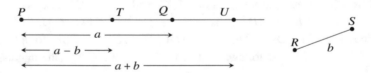

**Fig. 13.5** Addition and subtraction of line segments

The proof is quite long and will be presented as a sequence of constructions.

**Theorem 13.4** *An isoceles trapezoid can be constructed.*

**Proof** Let $H$ be any point on $c(Q, b)$. Construct $H'$ its reflection around $\overline{PQ}$. Denote the length of $\overline{HH'}$ by $h$ (Fig. 13.6).

Construct the circles $c(H, b)$, $c(Q, h)$. Let $K$ be a point of intersection of the circles and construct $K'$ the reflection of $K$ around $\overline{PQ}$ (Fig. 13.7).

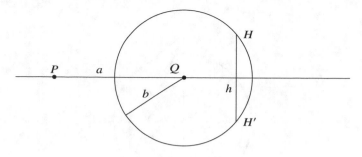

**Fig. 13.6** Construction of an isoceles trapezoid (1)

The line containing $\overline{PQ}$ is the perpendicular bisector of $\overline{HH'}$ and $\overline{KK'}$ so $\overline{HH'} \parallel \overline{KK'}$. $\overline{KH} = b$ since it is the radius of the circle centered on $H$, and $K', H'$ are reflections of $K, H$. $\triangle QQ'H \cong \triangle QQ'H'$ by side-side-side and $\triangle KQH \cong \triangle K'QH'$ by side-angle-side, so $\overline{K'H'} = \overline{KH} = b$. It follows that $\overline{KHH'K'}$ is an isosceles trapezoid whose bases are $\overline{HH'} = h$, $\overline{KK'} = 2h$ (Fig. 13.8). Denote the length of the diagonals $\overline{K'H} = \overline{KH'}$ by $d$. □

**Theorem 13.5** *An isoceles trapezoid can be circumscribed by a circle.*

**Proof** The theorem follows immediately from Thms. A.15 and A.16. □

**Theorem 13.6** *For $d, b, h$ shown in Fig. 13.8, $d^2 = b^2 + 2h^2$.*

**Proof** The theorem follows from Ptolemy's theorem (Thm. A.18) which says that in a quadrilateral circumscribed by a circle the product of the diagonals equals the sum of the products of the opposite sides. □

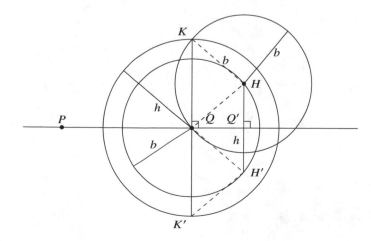

**Fig. 13.7** Construction of an isoceles trapezoid (2)

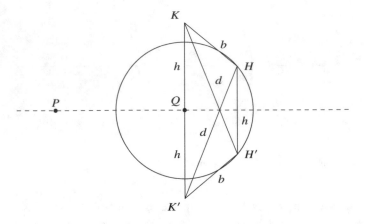

**Fig. 13.8** Construction of an isoceles trapezoid (3)

The proof of Thm. 13.3 can now be given.

***Proof*** Let $X$ be the point on line $\overline{PQ}$ that extends $\overline{PQ}$ by $b$. (We will eventually construct $X$.) Define $x = \overline{K'X}$. From Thm. 13.6:

$$d^2 = b^2 + 2h^2 = (x^2 - h^2) + 2h^2 = x^2 + h^2 .$$

Since $\triangle QK'X$ is a right triangle $x^2 = b^2 + h^2$ (Fig. 13.9).

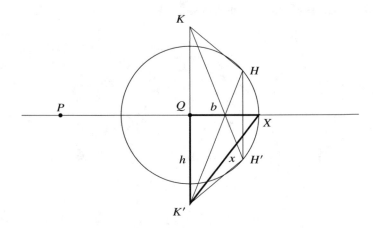

**Fig. 13.9** Application of Ptolemy's theorem

Construct $S$ as the intersection of $c(K, d), c(K', d)$ (Fig. 13.10). $\triangle QSK'$ is a right triangle so by Pythagoras's Theorem $\overline{QS}^2 = d^2 - h^2 = x^2$ and $\overline{QS} = x$.

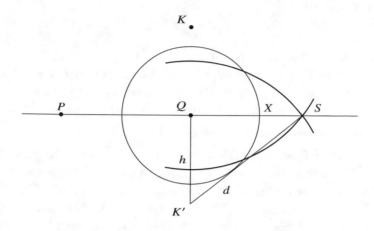

**Fig. 13.10**  Construction of the point for addition and subtraction (1)

Construct $X$ as the intersection of $c(K, x), c(K', x)$ (Fig. 13.11). Since the length of $\overline{QX}$ is $\sqrt{x^2 - h^2} = b$ the length of $\overline{PX}$ is $a + b$ and the length of $\overline{PX'}$ is $a - b$.  $\square$

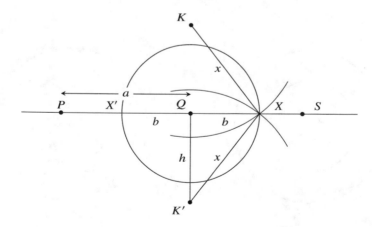

**Fig. 13.11**  Construction of the point for addition and subtraction (2)

## 13.5  Construction of a Line Segment as a Ratio of Segments

**Theorem 13.7** *Given line segments of length n, m, s, it is possible to construct a line segment of length:*

$$x = \frac{n}{m}s \, .$$

***Proof*** Construct two concentric circles $c_1 = c(Z, m)$ and $c_2 = c(Z, n)$,[1] and choose an arbitrary point $A$ on $c_1$. By Thm. 13.2 construct a chord $\overline{AB}$ of length $s$ on $c_1$ (Fig. 13.12a). If $\overline{AB}$ intersects $c_2$, by Thm. 13.3 multiply $m, n$ by a number $k$ so that the chord does not intersect the circle. Note that this does not change the value that we are trying to construct since $x = \dfrac{kn}{km}s = \dfrac{n}{m}s$.

Choose a point $H$ on $c_2$ and denote the length of $\overline{AH}$ by $w$. Construct $K$ on $c_2$ such that the length of $\overline{BK}$ is $w$ (Fig. 13.12b). $\triangle AHZ \cong \triangle BZK$ by side-side-side since $\overline{ZA} = \overline{ZB} = m$ are the radii of the same circle, as are $\overline{ZH} = \overline{ZK} = n$, and $\overline{AH} = \overline{BK} = w$ by construction (Fig. 13.13a). From $\triangle AHZ \cong \triangle BZK$ it follows $\angle AZH = \angle BZK$ and then $\angle AZB = \angle HZK$. It is difficult to see this equality from the diagram, but Fig. 13.13b should clarify the relation among the angles.

$\triangle ZAB \sim \triangle ZHK$ since both are isosceles triangles and we have shown that they have the same vertex angle. Label $\overline{HK}$ by $x$. Then:

$$\frac{m}{s} = \frac{n}{x}$$
$$x = \frac{n}{m}s \, . \qquad\qquad \square$$

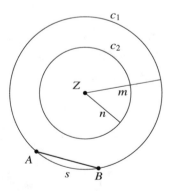

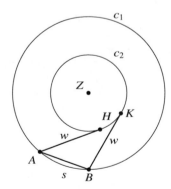

**Fig. 13.12a**  Construction of $x = \frac{n}{m}s$, step 1        **Fig. 13.12b**  Construction of $x = \frac{n}{m}s$, step 2

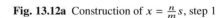

[1] We assume that $m > n$; if not, exchange the notation.

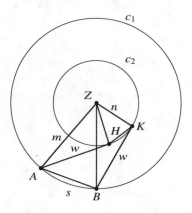

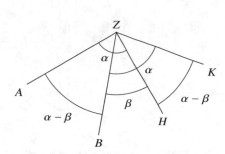

**Fig. 13.13a** Construction of $x = \frac{n}{m}s$, step 3     **Fig. 13.13b** $\angle AZB = \angle HZK$

## 13.6 Construction of the Intersection of Two Lines

**Theorem 13.8** *Given two lines containing the line segments $\overline{AB}, \overline{CD}$, it is possible to construct their intersection S.*

**Proof** Let $C', D'$ be the reflections of $C, D$ around $\overline{AB}$. There are two cases depending on whether $C, D$ lie on the same side of $\overline{AB}$ or on different sides. Label $x = \overline{CS}, c = \overline{CC'}, d = \overline{DD'}, e = \overline{CD}$ as shown in Figs. 13.14, 13.15. We compute the value of $x$ for each case.

    *Case 1:* $C, D$ are on the different sides of $\overline{AB}$. $S$ lies on $\overline{AB}$ because $\triangle CZS \cong \triangle C'ZS$ by side-angle-side: $\overline{CZ} = \overline{C'Z}, \angle CZS = \angle C'ZS = 90°$ and $\overline{ZS}$ is a common side. Therefore $\overline{C'S} = \overline{CS}$ and similarly $\overline{D'S} = \overline{DS}$. $\triangle CSC' \sim \triangle DSD'$ are similar so $\dfrac{x}{e-x} = \dfrac{c}{d}$ and solving the equation gives $x = \dfrac{c}{c+d}e$.

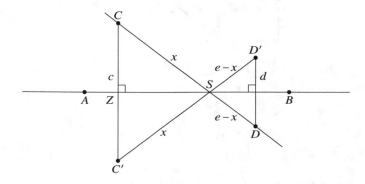

**Fig. 13.14** Construction of the intersection of two lines (1)

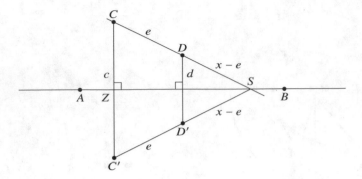

**Fig. 13.15** Construction of the intersection of two lines (2)

*Case 2:* $C, D$ are on the same side of $\overline{AB}$. $\triangle CSC' \sim \triangle DSD'$ gives $\dfrac{x}{x-e} = \dfrac{c}{d}$ and solving the equation gives $x = \dfrac{c}{c-d}e$.

Construct the circles $c(C', d), c(D, e)$ and denote their intersection by $H$ (Fig. 13.16). The sum of the line segments $\overline{CC'}, \overline{C'H}$ is $c + d$. We have to show that $H$ is on the extension of $\overline{CC'}$ so that $\overline{CH}$ is a line segment of length $c + d$. $\overline{CH} = c - d$ in case $D$ is on the same side of $\overline{AB}$ as $C$ (not shown in the diagram).

$H$ is the intersection of $c(C', d), c(D, e)$ so $\overline{DH} = e, \overline{C'H} = d$. By construction $\overline{C'D'} = e, \overline{D'D} = d$ so the quadrilateral $C'D'DH$ is a parallelogram.

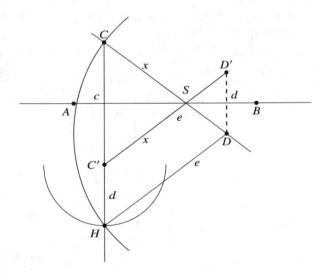

**Fig. 13.16** Construction of the intersection of two lines (3)

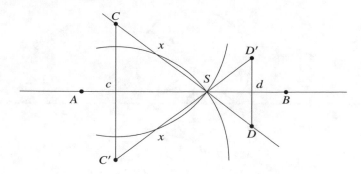

**Fig. 13.17** Construction of the intersection of two lines (4)

By construction $\overline{DD'} \parallel \overline{CC'}$ so $\overline{C'H} \parallel \overline{DD'}$ and therefore $\overline{C'H} \parallel \overline{CC'}$. Since one of its end points is $C'$ it must be on the line containing $\overline{CC'}$. By Thm. 13.3, from the lengths $c, d, e$ a line segment of length $c + d$ can be constructed and by Thm. 13.7 a line segment of length $x = \dfrac{c}{c + d} e$ can be constructed. $S$, the intersection of $c(C', x)$ and $c(C, x)$, is also the intersection of $\overline{AB}, \overline{CD}$ (Fig. 13.17).  □

## 13.7 Construction of the Intersection of a Line and a Circle

**Theorem 13.9** *Given a circle $k = C(M, r)$ and a line $l$ it is possible to construct the intersections of $k$ and $l$.*

**Proof** Construct $M'$, be the reflection of $M$ around $l$ and construct the circle $k' = c(M', r)$. Since $MYM' \cong \triangle MXM'$, $X, Y$, the points of intersection of $k, k'$, are the points of intersection of $l$ and $k$ (Fig. 13.18).

This construction cannot be done if $M$ is on the line $l$. In that case choose an arbitrary point $A$ on $l$ that is at a distance more than $r$ from $M$. Using Thm 13.3 shorten and lengthen $\overline{AM}$ by $r$. $X, Y$, the endpoints of these segments, are the intersections of $k$ and $l$ (Fig. 13.19).  □

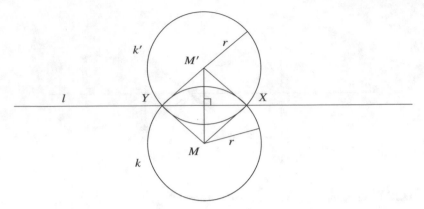

**Fig. 13.18** Construction of the intersection of a line and a circle (1)

## What Is the Surprise?

When one learns about constructions with a straightedge and compass it is obvious that both tools are necessary. Therefore, it was quite a surprise to find out that a compass is sufficient. The proof is quite long so we are not going to leave the straightedge at home, but the theorem shows that we should not assume that there are no alternatives to well-known mathematical concepts.

## Sources

This chapter is based on problem 33 of [13] reworked by Michael Woltermann [14]. An additional proof can be found in [25].

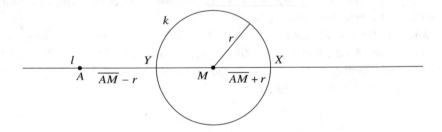

**Fig. 13.19** Construction of the intersection of a line and a circle (2)

# Chapter 14

# A Straightedge and One Circle is Sufficient

Can every construction with a straightedge and compass be done with only a straight-edge? The answer is no because lines are defined by linear equations and cannot represent circles which are defined by quadratic equations. In 1822 Jean-Victor Poncelet conjectured that a straightedge only is sufficient provided that *one circle* exists in the plane. This was proved in 1833 by Jakob Steiner.

After explaining in Sect. 14.1 what is meant by performing a construction with only a straightedge and one circle, the proof is presented in stages starting with five auxiliary constructions: construction of a line parallel to a given line (Sect. 14.2), construction of a perpendicular to a given line (Sect. 14.3), copying a line segment in a given direction (Sect. 14.4), construction of a line segment as a ratio of segments (Sect. 14.5) and construction of a square root (Sect. 14.6). Section 14.7 shows how to find the intersection(s) of a line with a circle and Sect. 14.8 shows how to find the intersection(s) of two circles.

## 14.1 What Is a Construction With Only a Straightedge?

A construction using a straightedge and compass is a sequence of three operations:

- Find the point of intersection of two lines.
- Find the point(s) of intersection of a line and a circle.
- Find the point(s) of intersection of two circles.

The first operation can be performed with a straightedge only.

A circle is defined by a point $O$, its *center*, and by a *radius* $r$, a line segment of length $r$ one of whose endpoints is the center. If we can construct the points labeled $X$ and $Y$ in Fig. 14.1a we can claim to have successfully constructed the points of intersection of a given circle with a given line. Similarly, the construction of $X, Y$ in Fig. 14.1b is the construction of the points of intersection of two given circles. The

© The Author(s) 2022

M. Ben-Ari, *Mathematical Surprises*, https://doi.org/10.1007/978-3-031-13566-8_14

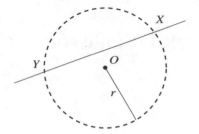 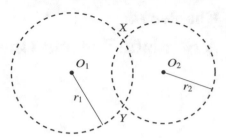

**Fig. 14.1a** $X, Y$ are the points of intersection of a line and a circle

**Fig. 14.1b** $X, Y$ are the points of intersections of two circles

circles drawn with dashed lines in a diagram do not actually appear in a construction; they are just used to help understand the construction.

The single given circle used in the constructions, called the *fixed circle*, can appear anywhere in the plane and can have an arbitrary radius.

## 14.2  Construction of a Line Parallel to a Given Line

**Theorem 14.1** *Given a line l defined by two points A, B and a point P not on the line, it is possible to construct a line through P that is parallel to $\overline{AB}$.*

**Proof**  There are two cases to the proof.

*Case 1:* $\overline{AB}$ is a *directed line segment* if the midpoint $M$ of $\overline{AB}$ is given. Construct a ray that extends $\overline{AP}$ and choose any point $S$ on the ray beyond $P$. Construct the lines $\overline{BP}, \overline{SM}, \overline{SB}$. The intersection of $\overline{BP}$ and $\overline{SM}$ is denoted $O$. Construct a ray that extends $\overline{AO}$ and denote by $Q$ the intersection of the ray $\overline{AO}$ with $\overline{SB}$ (Fig. 14.2).

We claim that $\overline{PQ} \parallel \overline{AB}$.

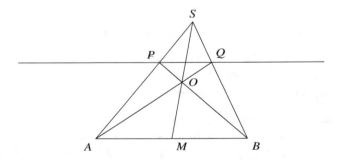

**Fig. 14.2** Construction of a parallel line in the case of a directed line

The proof uses Ceva's theorem.

*Ceva's theorem (Thm. A.19):* If the line segments from the vertices of a triangle to the opposite edges intersect in a point $O$ (as in Fig. 14.2), the lengths of the segments satisfy:

$$\frac{\overline{AM}}{\overline{MB}} \cdot \frac{\overline{BQ}}{\overline{QS}} \cdot \frac{\overline{SP}}{\overline{PA}} = 1 .$$

In Fig. 14.2 $M$ is the midpoint of $\overline{AB}$ so $\dfrac{\overline{AM}}{\overline{MB}} = 1$ and the equation becomes:

$$\frac{\overline{BQ}}{\overline{QS}} = \frac{\overline{PA}}{\overline{SP}} = \frac{\overline{AP}}{\overline{PS}} , \qquad (14.1)$$

since the order of the endpoints of a line segment is not important.

We claim that $\triangle ABS \sim \triangle PQS$:

$$\frac{\overline{BS}}{\overline{QS}} = \frac{\overline{BQ}}{\overline{QS}} + \frac{\overline{QS}}{\overline{QS}} = \frac{\overline{BQ}}{\overline{QS}} + 1$$

$$\frac{\overline{AS}}{\overline{PS}} = \frac{\overline{AP}}{\overline{PS}} + \frac{\overline{PS}}{\overline{PS}} = \frac{\overline{AP}}{\overline{PS}} + 1 .$$

Using Eq. 14.1:

$$\frac{\overline{BS}}{\overline{QS}} = \frac{\overline{BQ}}{\overline{QS}} + 1 = \frac{\overline{AP}}{\overline{PS}} + 1 = \frac{\overline{AP}}{\overline{PS}} + \frac{\overline{PS}}{\overline{PS}} = \frac{\overline{AS}}{\overline{PS}} ,$$

and it follows that $\triangle ABS \sim \triangle PQS$ and therefore $\overline{PQ} \parallel \overline{AB}$.

*Case 2:* $\overline{AB}$ is not necessarily a directed line segment. The fixed circle $c$ has center $O$ and radius $r$. $P$ is the point not on the line through which it is required to construct a line parallel to $l$ (Fig. 14.3a).

Choose $M$, any point on $l$, and construct a ray extending $\overline{MO}$ that intersects the circle at $U, V$. $\overline{UV}$ is a directed line segment because $O$, the center of the circle, bisects the diameter $\overline{UV}$. Choose a point $A$ on $l$ and use the construction for a directed line segment (Case 1) to construct a line through $A$ parallel to $\overline{UV}$ which intersects the circle at $X, Y$ (Fig. 14.3b).

Construct a diameter from $X$ through $O$ that intersects the other side of the circle at $X'$, and similarly construct the diameter $\overline{YY'}$. Construct the ray from $X'$ through $Y'$ and denote by $B$ its intersection with $l$. We claim that $M$ is the bisector of $\overline{AB}$ so that $\overline{AB}$ is a directed line segment and therefore a line can be constructed through $P$ parallel to $l$ (Fig. 14.4).

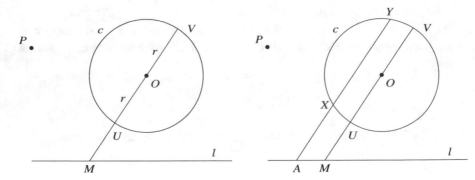

**Fig. 14.3a** Construction of a directed line

**Fig. 14.3b** Construction of a line parallel to the directed line

$\overline{OX}, \overline{OX'}, \overline{OY}, \overline{OY'}$ are all radii of the circle and $\angle XOY = \angle X'OY'$ since they are vertical angles, so $\triangle XOY \cong \triangle X'OY'$ by side-angle-side. Define[1] $l'$ to be a line through $O$ parallel to $l$ that intersects $\overline{XY}$ at $Z$ and $\overline{X'Y'}$ at $Z'$. $\angle XOZ = \angle X'OZ'$ are vertical angles, $\angle ZXO = \angle Z'X'O$ are alternate interior angles and $\overline{XO} = \overline{XO'}$ are radii, so $\triangle XOZ \cong \triangle X'OZ'$ by angle-side-angle and $\overline{ZO} = \overline{OZ'}$. Therefore, $\overline{AMOZ}$ and $\overline{BMOZ'}$ are parallelograms and $\overline{AM} = \overline{ZO} = \overline{OZ'} = \overline{MB}$.            □

**Theorem 14.2** *Given a line segment $\overline{AB}$ and a point $P$ not on the line, it is possible to construct a line segment $\overline{PQ}$ that is parallel to $\overline{AB}$ and whose length is equal to the length of $\overline{AB}$, that is, it is possible to copy $\overline{AB}$ parallel to itself with $P$ as one of its endpoints.*

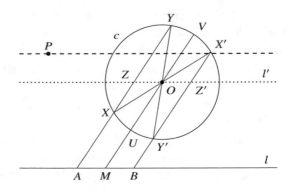

**Fig. 14.4** Proof that $l'$ is parallel to $l$

---

[1] Define, not construct, because we are in the middle of the proof that such a line can be constructed.

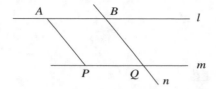

**Fig. 14.5** Construction of a copy of a line parallel to an existing line

**Proof** We have proved that it is possible to construct a line $m$ through $P$ parallel to $\overline{AB}$ and a line $n$ through $B$ to parallel to $\overline{AP}$. The quadrilateral $\overline{ABQP}$ is a parallelogram so opposite sides are equal $\overline{AB} = \overline{PQ}$ (Fig. 14.5).                                    □

## 14.3 Construction of a Perpendicular to a Given Line

**Theorem 14.3** *Given a line segment l and a point P not on l, it is possible to construct a perpendicular to l through P.*

**Proof** By Thm. 14.1 construct a line $l'$ parallel to $l$ that intersects the fixed circle at $U, V$. Construct the diameter $\overline{UOU'}$ and the chord $\overline{VU'}$ (Fig. 14.6). $\angle UVU'$ is a right angle because it is subtended by a diameter. Therefore $\overline{VU'}$ is perpendicular to $\overline{UV}$ and $l$. Again by Thm. 14.1 construct the parallel to $\overline{VU'}$ through $P$.                                    □

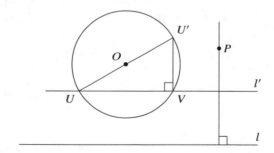

**Fig. 14.6** Construction of a perpendicular line

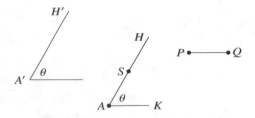

**Fig. 14.7** Copying a line segment in a given direction

## 14.4 Copying a Line Segment in a Given Direction

**Theorem 14.4** *It is possible to construct a copy of a given line segment in the direction of another line.*

The meaning of "direction" is that the line defined by two points $A', H'$ is at an angle $\theta$ relative to some axis and the goal is to construct $\overline{AS} = \overline{PQ}$ such that $\overline{AS}$ will have the same angle $\theta$ relative to that axis (Fig. 14.7).

***Proof*** By Thm. 14.1 it is possible to construct a line segment $\overline{AH}$ such that $\overline{AH} \parallel \overline{A'H'}$, and to construct a line segment $\overline{AK}$ such that $\overline{AK} \parallel \overline{PQ}$. $\angle HAK = \theta$ so it remains to find a point $S$ on $\overline{AH}$ so that $\overline{AS} = \overline{PQ}$.

Construct two radii $\overline{OU}, \overline{OV}$ of the fixed circle which are parallel to $\overline{AH}, \overline{AK}$, respectively, and construct a ray through $K$ parallel to $\overline{UV}$. Denote its intersection with $\overline{AH}$ by $S$ (Fig. 14.8). By construction, $\overline{AH} \parallel \overline{OU}$ and $\overline{AK} \parallel \overline{OV}$, so $\angle SAK = \angle HAK = \angle UOV = \theta$. $\overline{SK} \parallel \overline{UV}$ and $\triangle SAK \sim \triangle UOV$ by angle-angle-angle, $\triangle UOV$ is isosceles because $\overline{OU}, \overline{OV}$ are radii of the same circle. Therefore, $\triangle SAK$ is isosceles and $\overline{AS} = \overline{AK} = \overline{PQ}$. $\qquad\square$

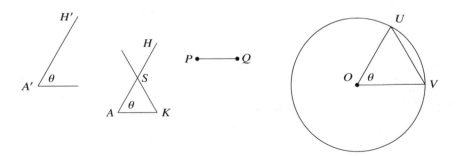

**Fig. 14.8** Using the fixed circle to copy the line segment

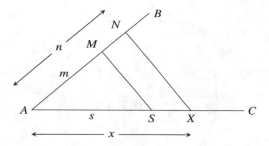

**Fig. 14.9** Similar triangles to construct the ratio of lengths

## 14.5  Construction of a Line Segment as a Ratio of Segments

**Theorem 14.5** *Given line segments of lengths* $n, m, s$, *it is possible to construct a line segment of length:*

$$x = \frac{n}{m}s\,.$$

**Proof** Choose points $A, B, C$ not on the same line and construct rays $\overrightarrow{AB}, \overrightarrow{AC}$. By Thm. 14.4 it is possible to construct points $M, N, S$ such that $\overline{AM} = m$, $\overline{AN} = n$, $\overline{AS} = s$. By Thm. 14.1 construct a line through $N$ parallel to $\overline{MS}$ which intersects $\overline{AC}$ at $X$ and label $\overline{AX}$ by $x$ (Fig. 14.9). $\triangle MAS \sim \triangle NAX$ by angle-angle-angle so $\frac{m}{n} = \frac{s}{x}$ and $x = \frac{n}{m}s$.                                                                 □

## 14.6  Construction of a Square Root

**Theorem 14.6** *Given line segments of lengths* $a, b$, *it is possible to construct a line segment of length* $\sqrt{ab}$.

**Proof** We want to express $x = \sqrt{ab}$ as $x = \frac{n}{m}s$ in order to use Thm. 14.5.

- For $n$ we use $d$, the diameter of the fixed circle.
- For $m$ we use $t = a + b$ which can be constructed from $a, b$ by Thm. 14.4.
- We define $s = \sqrt{hk}$ where $h, k$ are defined as expressions on the lengths $a, b, t, d$.

Define $h = \frac{d}{t}a$ and $k = \frac{d}{t}b$ and then compute:

$$x = \sqrt{ab} = \sqrt{\frac{th}{d}\frac{tk}{d}} = \sqrt{\left(\frac{t}{d}\right)^2 hk} = \frac{t}{d}\sqrt{hk} = \frac{t}{d}s$$

$$h + k = \frac{d}{t}a + \frac{d}{t}b = \frac{d(a+b)}{t} = \frac{dt}{t} = d\,.$$

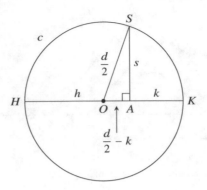

**Fig. 14.10** Construction of a square root

By Thm. 14.4 construct $\overline{HA} = h$ on a diameter $\overline{HK}$ of the fixed circle. From $h + k = d$ we have $\overline{AK} = k$ (Fig. 14.10). By Thm. 14.3 construct a perpendicular to $\overline{HK}$ at $A$ and denote the intersection of this line with the circle by $S$. $\overline{OS} = \overline{OK} = d/2$ and $\overline{OA} = (d/2) - k$.

By Pythagoras's Theorem:

$$s^2 = \left(\frac{d}{2}\right)^2 - \left(\frac{d}{2} - k\right)^2$$
$$= \left(\frac{d}{2}\right)^2 - \left(\frac{d}{2}\right)^2 + 2\frac{dk}{2} - k^2$$
$$= k(d - k) = kh$$
$$s = \sqrt{hk}.$$

Now $x = \frac{t}{d}s$ can be constructed by Thm. 14.5.                                   □

## 14.7  Construction of the Intersection of a Line and a Circle

**Theorem 14.7** *Given a line l and a circle $c(O, r)$, it is possible to construct their points of intersection (Fig. 14.11).*

**Proof** By Thm. 14.3 it is possible to construct a perpendicular from the center of the circle $O$ to the line $l$. The intersection of $l$ with the perpendicular is denoted by $M$. $\overline{OM}$ bisects the chord $\overline{XY}$, where $X, Y$ are the intersections of the line with the circle (Fig. 14.12). Define $\overline{XY} = 2s$ and $\overline{OM} = t$. Note that $s, X, Y$ are just definitions not entities have been constructed.

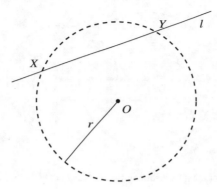

**Fig. 14.11** Construction of the points of intersection of a line and a circle (1)

By Pythagoras's Theorem $s^2 = r^2 - t^2 = (r+t)(r-t)$. By Thm. 14.4 it is possible to construct line segments of length $t$ from $O$ in the two directions $\overline{OR}$ and $\overline{RO}$. The result is two line segments of length $r + t, r - t$.

By Thm. 14.6 a line segment of length $s = \sqrt{(r+t)(r-t)}$ can be constructed, and by Thm. 14.4 line segments of length $s$ from $M$ along $l$ in both directions can be constructed. Their other endpoints are the points of intersection of $l$ and $c$.     □

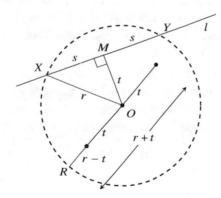

**Fig. 14.12** Construction of the points of intersection of a line and a circle (2)

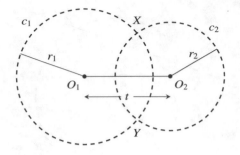

**Fig. 14.13** Construction of the intersection of two circles (1)

## 14.8 Construction of the Intersection of Two Circles

**Theorem 14.8** *Given two circles* $c(O_1, r_1), c(O_2, r_2)$, *it is possible to construct their points of intersection.*

***Proof*** Construct $\overline{O_1 O_2}$ and label its length $t$ (Fig. 14.13). Label by $A$ be the point of intersection of $\overline{O_1 O_2}$ and $\overline{XY}$, and label $q = \overline{O_1 A}$, $x = \overline{XA}$ (Fig. 14.14). $A$ has not yet been constructed, but if $q, x$ are constructed then by Thm. 14.4 the point $A$ at length $q$ from $O_1$ in the direction $\overline{O_1 O_2}$ can be constructed.

Once $A$ has been constructed, by Thm. 14.3 a perpendicular to $\overline{O_1 O_2}$ at $A$ can be constructed, and by Thm. 14.4 it is possible to construct line segments of length $x$ from $A$ in both directions along the perpendicular. Their other endpoints are the points of intersection of the circles.

**Construction of the length** $q$: Define $d = \sqrt{r_1^2 + t^2}$, the hypotenuse of a right triangle, which can be constructed from the known lengths $r_1, t$. Note that $\triangle O_1 X O_2$ is not necessarily a right triangle; the right triangle can be constructed anywhere in the plane. In the right triangle $\triangle X A O_1$, $\cos \angle X O_1 A = q/r_1$. By the Law of Cosines for $\triangle X O_1 O_2$:

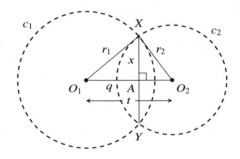

**Fig. 14.14** Construction of the intersection of two circles (2)

$$r_2^2 = t^2 + r_1^2 - 2r_1 t \cos \angle XO_1O_2$$
$$= t^2 + r_1^2 - 2tq$$
$$2tq = (t^2 + r_1^2) - r_2^2 = d^2 - r_2^2$$
$$q = \frac{(d+r_2)(d-r_2)}{2t} \, .$$

By Thm. 14.4 these lengths can be constructed and by Thm. 14.5 $q$ can be constructed from $d + r_2, d - r_2, 2t$.

**Construction of the length $x$:** By Pythagoras's Theorem:

$$x = \sqrt{r_1^2 - q^2} = \sqrt{(r_1 + q)(r_1 - q)} \, .$$

By Thm. 14.4, $h = r_1 + q, k = r_1 - q$ can be constructed, as can $x = \sqrt{hk}$ by Thm. 14.6.                                                                                         $\square$

## What Is the Surprise?

A compass is necessary because a straightedge can only compute the roots of linear equations and not values such as $\sqrt{2}$, the hypotenuse of an isoceles right-triangle with sides of length 1. However, it is surprising that the existence of only one circle, regardless of the position of its center and the length of its radius, is sufficient to perform any construction that is possible with a straightedge and compass.

## Sources

This chapter is based on problem 34 of [13] reworked by Michael Woltermann [14].

# Chapter 15

# Are Triangles with Equal Areas and Perimeters Congruent?

Are two triangles with the same area and the same perimeter congruent? Not necessarily: the triangles with sides $(17, 25, 28)$ and $(20, 21, 29)$ both have perimeter 70 and area 210 but they are not congruent (Fig. 15.1).[1] This chapter shows that given a triangle with rational sides it is possible to construct a non-congruent triangle, also with rational sides, that has the same area and the perimeter. We carry out the derivation using an example, showing that the triangle with sides $(3, 4, 5)$ and the triangle with sides $\left(\frac{156}{35}, \frac{101}{21}, \frac{41}{15}\right)$ both have perimeter 12 and area 6.

## 15.1 From a Triangle to an Elliptic Curve

The three angle bisectors in a triangle intersect in a point called the *incenter* of the triangle. The incenter is the center of a circle inscribed within the triangle (Fig. 15.2).

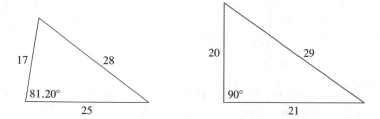

**Fig. 15.1** Non-congruent triangles with the same area and the same perimeter

---

[1] The areas were computed using Heron's formula (Thm. A.3) and the angles using the Law of Cosines (Thm. A.8).

M. Ben-Ari, *Mathematical Surprises*, https://doi.org/10.1007/978-3-031-13566-8_15

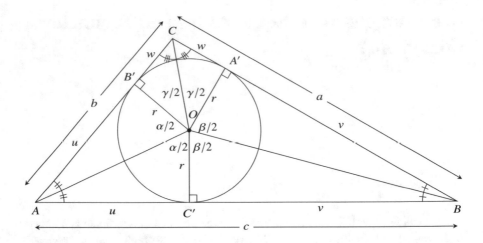

**Fig. 15.2** A circle inscribed within a triangle

Drop altitudes from the center $O$ to the sides. The altitudes have length $r$, the radius of the inscribed circle. The altitudes and angle bisectors create three pairs of congruent right triangles:

$$\triangle AOB' \cong \triangle AOC', \quad \triangle BOA' \cong \triangle BOC', \quad \triangle COA' \cong \triangle COB'.$$

The altitudes divide the sides $a, b, c$ into segments $u, v, w$. The area of $\triangle ABC$ is the sum of the areas of $\triangle BOC, \triangle AOB, \triangle AOC$:

$$A = \frac{1}{2}(w + v)r + \frac{1}{2}(v + u)r + \frac{1}{2}(u + w)r \tag{15.1a}$$

$$= \frac{1}{2} \cdot 2(u + v + w)r \tag{15.1b}$$

$$= \frac{1}{2}(a + b + c)r \tag{15.1c}$$

$$= sr, \tag{15.1d}$$

where $s$ is the *semi-perimeter*, one-half the perimeter of the triangle $\triangle ABC$. The lengths of $u, v, w$ can be expressed using the radius of the circle and the central angles $\alpha/2, \beta/2, \gamma/2$:

$$\tan\frac{\alpha}{2} = \frac{u}{r}, \quad \tan\frac{\beta}{2} = \frac{v}{r}, \quad \tan\frac{\gamma}{2} = \frac{w}{r}. \tag{15.2}$$

The semi-perimeter can now be expressed in terms of the tangents:

$$s = u + v + w = r \tan \frac{\alpha}{2} + r \tan \frac{\beta}{2} + r \tan \frac{\gamma}{2} = r \left( \tan \frac{\alpha}{2} + \tan \frac{\beta}{2} + \tan \frac{\gamma}{2} \right),$$

and by Eq. 15.1d the area is:

$$A = sr = r^2 \left( \tan \frac{\alpha}{2} + \tan \frac{\beta}{2} + \tan \frac{\gamma}{2} \right). \tag{15.3}$$

From $r = A/s$, Eq. 15.3 can be written as:

$$\tan \frac{\alpha}{2} + \tan \frac{\beta}{2} + \tan \frac{\gamma}{2} = \frac{A}{r^2} = \frac{A}{(A/s)^2} = \frac{s^2}{A}. \tag{15.4}$$

Since the sum of the angles $\alpha, \beta, \gamma$ is 360°:

$$\gamma/2 = 360°/2 - (\alpha/2 + \beta/2) \tag{15.5a}$$

$$\tan \gamma/2 = \tan(180° - (\alpha/2 + \beta/2)) \tag{15.5b}$$

$$= -\tan(\alpha/2 + \beta/2) \tag{15.5c}$$

$$= \frac{\tan \alpha/2 + \tan \beta/2}{\tan \alpha/2 \tan \beta/2 - 1}, \tag{15.5d}$$

using the formula for the tangent of the sum of two angles (Thm. A.9).

Let us simplify the notation by defining variables for the tangents:

$$x = \tan \frac{\alpha}{2}, \quad y = \tan \frac{\beta}{2}, \quad z = \tan \frac{\gamma}{2}. \tag{15.6}$$

By Eq. 15.5d we can express $z = \tan \gamma/2$ in terms of $x, y$:

$$z = \frac{x + y}{xy - 1}. \tag{15.7}$$

With this notation, Eq. 15.4 becomes:

$$x + y + \frac{x + y}{xy - 1} = \frac{s^2}{A}. \tag{15.8}$$

Given fixed values of $A$ and $s$ are there multiple solutions of Eq. 15.8?

For the right triangle $(3, 4, 5)$:

$$\frac{s^2}{A} = \frac{\left( \frac{1}{2}(3 + 4 + 5) \right)^2}{\frac{1}{2} \cdot 3 \cdot 4} = \frac{6^2}{6} = 6. \tag{15.9}$$

If there is another solution Eq. 15.8 with $s^2/A = 6$, it can be written as:

$$x + y + \frac{x+y}{xy-1} = 6 \tag{15.10a}$$

$$x^2 y + xy^2 - 6xy + 6 = 0. \tag{15.10b}$$

This is an equation for an *elliptic curve*.

## 15.2  Solving the Equation for the Elliptic Curve

A portion of the graph of Eq. 15.10b is shown Fig. 15.3. Any point on the closed curve in the first quadrant is a solution to the equation because the lengths of the sides of the triangle must be positive. $A, B, D$ correspond to the triangle $(3, 4, 5)$ as shown below. To find additional rational solutions the *method of two secants* is used.

Construct a secant through the points $A = (2, 3)$, $B = (1, 2)$. It intersects the curve at $C = (-1.5, -0.5)$, but this does not give a solution because the values are negative. Construct a second secant from $C$ to $D = (3, 2)$. The intersection with the curve at $E \approx (1.5, 1.2)$ does give a new solution whose coordinates will be computed below.

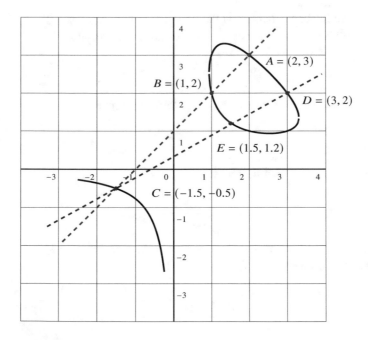

**Fig. 15.3** The method of two secants

The equation of the (red) line through $A, B$ is $y = x + 1$. From Eq. 15.10b:

$$x^2(x+1) + x(x+1)^2 - 6x(x+1) + 6 = 0$$
$$2x^3 - 3x^2 - 5x + 6 = 0 .$$

From $A, B$ we know two roots $x = 2, x = 1$ so we can factor the cubic polynomial:

$$(x-2)(x-1)(ax+b) = 0 ,$$

where the third root is unknown. Multiply the factors and conclude that $a = 2, b = 3$ since $2x^3 - 3x^2 - 5x + 6 = ax^3 + \cdots + 2b$. The third factor is $2x + 3$ which gives the third root $x = -\frac{3}{2}$ and $y = x + 1 = -\frac{1}{2}$. This is the point $C = (-\frac{3}{2}, -\frac{1}{2})$ in the graph.

The equation of the (blue) line through $C, D$ is:

$$y = \frac{5}{9}x + \frac{1}{3} . \tag{15.11}$$

Substitute for $y$ in Eq. 15.10b:

$$x^2\left(\frac{5}{9}x + \frac{1}{3}\right) + x\left(\frac{5}{9}x + \frac{1}{3}\right)^2 - 6x\left(\frac{5}{9}x + \frac{1}{3}\right) + 6 = 0$$
$$\frac{70}{81}x^3 - \frac{71}{27}x^2 - \frac{17}{9}x + 6 = 0 .$$

From $C, D$ we know two roots $x = 3, x = -\frac{3}{2}$ so we can factor the cubic polynomial:

$$(x-3)\left(x + \frac{3}{2}\right)(ax+b) = 0 .$$

Equating the coefficients of the cubic term and the constant terms gives:

$$\frac{70}{81}x - \frac{4}{3} = 0$$
$$x = \frac{54}{35} \approx 1.543 ,$$

and $y$ can be computed from Eq. 15.11:

$$y = \frac{25}{21} \approx 1.190 .$$

The coordinates of $E$ are:

$$\left(\frac{54}{35}, \frac{25}{21}\right) = (1.543, 1.190) ,$$

which are close to the approximations $(1.5, 1.2)$ obtained from the graph.

Finally, compute $z$ from Eq. 15.7:

$$z = \frac{x+y}{xy-1} = \left(\frac{54}{35} + \frac{25}{21}\right) \Big/ \left(\frac{54}{35}\frac{25}{21} - 1\right) = \frac{2009}{615} = \frac{49}{15}.$$

## 15.3 Derivation of a Triangle From the Elliptic Curve

Using Eqs. 15.2, 15.6, $a, b, c$, the sides of the triangle $\triangle ABC$, can be computed from $x, y, z$ and $r = A/s = 6/6 = 1$:

$$a = w + v = r(z + y) = (z + y)$$
$$b = u + w = r(x + z) = (x + z)$$
$$c = u + v = r(x + y) = (x + y).$$

For solution $A$ of the elliptic curve the sides of the triangle are:

$$a = z + y = 1 + 3 = 4$$
$$b = x + z = 2 + 1 = 3$$
$$c = x + y = 2 + 3 = 5.$$

For solution $E$ of the elliptic curve the sides of the triangle are:

$$a = z + y = \frac{49}{15} + \frac{25}{21} = \frac{156}{35}$$

$$b = x + z = \frac{54}{35} + \frac{49}{15} = \frac{101}{21}$$

$$c = x + y = \frac{54}{35} + \frac{25}{21} = \frac{41}{15}.$$

Let us check this result. The semi-perimeter is:

$$s = \frac{1}{2}\left(\frac{156}{35} + \frac{101}{21} + \frac{41}{15}\right) = \frac{1}{2}\left(\frac{468 + 505 + 287}{105}\right) = \frac{1}{2}\left(\frac{1260}{105}\right) = 6,$$

and the area can be computed using Heron's formula (Thm. A.3):

$$A = \sqrt{6\left(6 - \frac{156}{35}\right)\left(6 - \frac{101}{21}\right)\left(6 - \frac{41}{15}\right)} = \sqrt{36} = 6.$$

Is $\left(\frac{156}{35}, \frac{101}{21}, \frac{41}{15}\right) \cong (3, 4, 5)$? To simplify the computation let us use the decimal approximations $(4.48, 4.81, 2.73)$. Then:

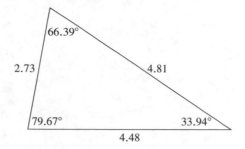

**Fig. 15.4** The triangle with the same perimeter and area as $(3, 4, 5)$

$$\sqrt{4.48^2 + 2.73^2} = 5.25 \neq 4.81\,,$$

so this is not a right triangle and not congruent to $(3, 4, 5)$.

The Law of Cosines can be used to compute the angles of the triangle as shown in Fig. 15.4.

## What Is the Surprise?

Are triangles with the same area and perimeter congruent? My first impression was to say "yes" because it is not easy to find counterexamples. What is surprising is that given an arbitrary triangle with rational sides, it is possible to construct a non-congruent triangle with rational sides which has the same area and perimeter, although the result can be strange as with the triangles $(3, 4, 5)$ and $\left(\frac{156}{35}, \frac{101}{21}, \frac{41}{15}\right)$.

## Sources

This chapter is based on [33]. In [3] it is shown that given an isoceles triangle there are non-congruent triangles with the same area and perimeter, but the proof does not include an explicit construction.

# Chapter 16
# Construction of a Regular Heptadecagon

The only regular polygons that the Greeks knew how to construct with a straightedge and compass were the triangle, the square, the pentagon and the regular polygon with 15 sides. Given a regular polygon with $n$ sides, a polygon with $2n$ sides can be constructed by circumscribing the polygon with a circle and bisecting the central angle (Fig. 16.1). No further progress was made until 1796 when Carl Friedrich Gauss awoke one morning, just before his 19th birthday, and by "concentrated thought" figured out how to construct a regular *heptadecagon*, a regular polygon with 17 sides. This achievement inspired him to become a mathematician.

Section 16.1 discusses the relation between the side of a polygon inscribed in a circle and the central angle that it subtends. Section 16.2 states without proof the Fundamental Theorem of Algebra. Section 16.3 presents the *roots of unity*, the roots of the polynomial $x^n - 1$, which are central to Gauss's proof. Sections 16.4 and 16.5 present Gauss's proof which is based on symmetries of roots of polynomials. Gauss derived a *formula* proving that the heptadecagon is constructible, but a geometric construction was not given for almost a century. Section 16.6 gives an elegant construction by James J. Callagy. Section 16.7 shows how constructions of a regular pentagon can be derived using both geometry and trigonometry.

Some of the material is more straightforward if presented using complex numbers. This material is set off in boxes that can be skipped.

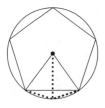

**Fig. 16.1** Constructing a regular polynomial with 10 sides from a regular pentagon

M. Ben-Ari, *Mathematical Surprises*, https://doi.org/10.1007/978-3-031-13566-8_16

## 16.1  Construction of Regular Polygons

The construction of the regular heptadecagon led to the Gauss-Wantzel theorem, which states that a regular polygon with $n$ sides can be constructed with a straightedge and compass if and only if $n$ is the product of a power of 2 and zero or more *distinct* Fermat numbers $2^{2^k} + 1$ which are prime. The known Fermat primes are:

$$F_0 = 3, \quad F_1 = 5, \quad F_2 = 17, \quad F_3 = 257, \quad F_4 = 65537 .$$

A regular polygon with 257 sides was constructed by Magnus Georg Paucker in 1822 and by Friedrich Julius Richelot 1832. In 1894 Johann Gustav Hermes claimed to have constructed a regular polygon with 65537 sides.

To construct a regular polygon it is sufficient to construct a line segment of length $\cos \theta$, where $\theta$ is the central angle subtended by a chord that is a side of the polygon inscribed in a unit circle. Given the line segment $\overline{OB} = \cos \theta$, construct a perpendicular at $B$ and label its intersection with the unit circle by $C$. Then:

$$\cos \theta = \frac{\overline{OB}}{\overline{OC}} = \overline{OB}$$
$$\theta = \cos^{-1}(\overline{OB}) .$$

The chord $\overline{AC}$ is a side of the regular polygon (Fig. 16.2).

Given a line segment defined to have length 1, the lengths that are constructible are those which can be obtained from line segments of known length using the operations $\{+, -, \times, /, \sqrt{}\}$ (Sect. 2.5). Gauss showed that $\cos(360°/17)$, the cosine of the central angle of a heptadecagon, is constructible since it can be expressed using only these operations:

$$\cos \left( \frac{360°}{17} \right) = -\frac{1}{16} + \frac{1}{16}\sqrt{17} + \frac{1}{16}\sqrt{34 - 2\sqrt{17}} +$$

$$\frac{1}{8}\sqrt{17 + 3\sqrt{17} - \sqrt{34 - 2\sqrt{17}} - 2\sqrt{34 + 2\sqrt{17}}} .$$

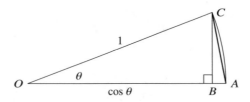

**Fig. 16.2** The cosine of the central angle of a regular polygon

## 16.2 The Fundamental Theorem of Algebra

The following theorem will be used without proof.

**Theorem 16.1** *Every polynomial of degree n has exactly n roots.*

The statement of the theorem has been simplified because all we will need to know is that *n* roots *exist*.

> **The Fundamental Theorem of Algebra** states that every non-constant polynomial of degree $n$ in a single variable with *complex* coefficients has exactly $n$ *complex* roots. If there are multiple roots with the same value, they are all counted: $x^2 - 4x + 4 = (x - 2)(x - 2)$ has two roots both equal to 2. The polynomial $x^2 + 1$ with integer coefficients has two complex roots $\pm\sqrt{-1}$. Strangely, even though the theorem is about finite algebraic entities—polynomials of degree $n$ with $n$ roots—methods of analysis, usually complex analysis, are needed to prove the theorem.

## 16.3 Roots of Unity

By the Fundamental Theorem of Algebra (Thm. 16.1) the polynomial $x^n - 1$ has $n$ roots for any integer $n > 1$. One root is $x = 1$ so there are $n - 1$ other roots. Denote one of these roots by $r$. Since $r^n = 1$ it is called an *n-th root of unity*. What about $r^2$?

$$(r^2)^n = (r^n)^2 = 1^2 = 1.$$

It follows that the $n$ numbers:

$$1, r, r^2, \ldots, r^{n-2}, r^{n-1}$$

are *n*-th roots of unity.

> Let $r = \cos\left(\frac{2\pi}{n}\right) + i \sin\left(\frac{2\pi}{n}\right)$. By de Moivre's formula:
>
> $$\left[\cos\left(\frac{2\pi}{n}\right) + i \sin\left(\frac{2\pi}{n}\right)\right]^n = \cos\left(\frac{2n\pi}{n}\right) + i \sin\left(\frac{2n\pi}{n}\right) = 1.$$

**Theorem 16.2** *Let n be a* prime number *and let r an n-th root of unity. Then:*

$$\{1, r, r^2, \ldots, r^{n-2}, r^{n-1}\}$$

*are distinct so they are* all *the n-th roots of unity.*

**Proof** Suppose that the powers are not distinct so that $r^i = r^j$ for some $0 \leq i < j \leq n-1$. Then $r^j/r^i = r^{j-i} = 1$ so there exists at least one positive integer $i'$ less than $n$ such that $r^{i'} = 1$. Let $m$ be the smallest such positive integer. By the division algorithm for integers $n = ml + k$ for some $0 < l < n$ and $0 \leq k < m$. From:

$$1 = r^n = r^{ml+k} = (r^m)^l \cdot r^k = 1^l \cdot r^k = r^k,$$

we have $0 \leq k < m$ and $r^k = 1$. Since $m$ was defined to be the smallest such positive integer $k = 0$ and $n = ml$ is not prime.                                                                   □

**Theorem 16.3** *Let $\{a_1, a_2, \ldots, a_{n-1}, a_n\}$ be the roots of an n-th degree polynomial $f(x)$. Then:*

$$f(x) = (x - a_1)(x - a_2) \cdots (x - a_{n-1})(x - a_n). \tag{16.1}$$

**Proof** If $a_i$ is a root of $f(x)$ by definition $f(a_i) = 0$ but:

$$f(a_i) = (a_i - a_1)(a_i - a_2) \cdots (a_i - a_{n-1})(a_i - a_n)$$
$$= \cdots (a_i - a_i) \cdots = 0.$$

Therefore, $f(x) = (x - a_i)g_i(x)$ for some $g_i(x)$ and by induction this holds for all the roots.                                                                   □

From Eq. 16.1 it is easy to see that the coefficient of $x^{n-1}$ is:

$$-(a_1 + a_2 + \cdots + a_{n-1} + a_n).$$

Since the coefficient of $x^{n-1}$ in $x^n - 1$ for $n \geq 2$ is zero, we have:

$$-(1 + r + r^2 + \cdots + r^{n-2} + r^{n-1}) = 0$$
$$r + r^2 + \cdots + r^{n-2} + r^{n-1} = -1.$$

For the heptadecagon this is:

$$r + r^2 + r^3 + r^4 + r^5 + r^6 + r^7 + r^8 +$$
$$r^9 + r^{10} + r^{11} + r^{12} + r^{13} + r^{14} + r^{15} + r^{16} = -1. \tag{16.2}$$

## 16.4 Gauss's Proof That a Heptadecagon Is Constructible

What Gauss understood is that one need not work with the roots in their natural order $r, r^2, \ldots, r^{16}$. The powers of $r^3$ give all the roots but in a different order:

$$r^1, \ r^{1\cdot3=3}, \ r^{3\cdot3=9}, \ r^{9\cdot3=27=10}, \ r^{10\cdot3=30=13}, \ r^{13\cdot3=39=5}, \ r^{5\cdot3=15}, \ r^{15\cdot3=45=11},$$

$$r^{11\cdot3=33=16}, \ r^{16\cdot3=48=14}, \ r^{14\cdot3=42=8}, \ r^{8\cdot3=24=7}, \ r^{7\cdot3=21=4}, \ r^{4\cdot3=12}, \ r^{12\cdot3=36=2}, \ r^{2\cdot3=6},$$

where the roots have been reduced modulo 17:

$$r^{17m+k} = (r^{17})^m \cdot r^k = 1^m \cdot r^k = r^k.$$

Check that the list contains all the roots (except 1) exactly once:

$$r^1, r^3, r^9, r^{10}, r^{13}, r^5, r^{15}, r^{11}, r^{16}, r^{14}, r^8, r^7, r^4, r^{12}, r^2, r^6. \qquad (16.3)$$

Given a monic quadratic polynomial whose roots are $a, b$:

$$y^2 + py + q = (y - a)(y - b) = 0,$$

we can compute the coefficients $p, q$ from the roots (Chap. 7):

$$p = -(a + b), \quad q = ab.$$

Therefore, *given* $a + b$ and $ab$ we can write down the quadratic equation of which $a, b$ are the roots.

Let $a_0$ be the sum of the roots in the odd positions in Eq. 16.3:

$$a_0 = r + r^9 + r^{13} + r^{15} + r^{16} + r^8 + r^4 + r^2,$$

and let $a_1$ be the sum of the roots in the even positions in Eq. 16.3:

$$a_1 = r^3 + r^{10} + r^5 + r^{11} + r^{14} + r^7 + r^{12} + r^6.$$

To obtain $a_0, a_1$ as roots of a quadratic equation first compute their sum and use Eq. 16.2:

$$a_0 + a_1 = r + r^2 + \cdots + r^{16} = -1.$$

Now we have to work very hard to compute their product. Figure 16.3 shows the computation where the values of $r^i r^j = r^{i+j}$ are written after reducing the exponents modulo 17. Check that each root occurs exactly four times so that—again using Eq. 16.2—the value of the product is −4.

$$a_0 a_1 = (r + r^9 + r^{13} + r^{15} + r^{16} + r^8 + r^4 + r^2) \times$$

$$(r^3 + r^{10} + r^5 + r^{11} + r^{14} + r^7 + r^{12} + r^6)$$

$$= \underset{1}{r^4} + \underset{1}{r^{11}} + \underset{1}{r^6} + \underset{1}{r^{12}} + \underset{1}{r^{15}} + \underset{1}{r^8} + \underset{1}{r^{13}} + \underset{1}{r^7} +$$

$$\underset{2}{r^{12}} + \underset{1}{r^2} + \underset{1}{r^{14}} + \underset{1}{r^3} + \underset{2}{r^6} + \underset{1}{r^{16}} + \underset{2}{r^4} + \underset{2}{r^{15}} +$$

$$\underset{2}{r^{16}} + \underset{3}{r^6} + \underset{1}{r^1} + \underset{2}{r^7} + \underset{1}{r^{10}} + \underset{2}{r^3} + \underset{2}{r^8} + \underset{2}{r^2} +$$

$$\underset{2}{r^1} + \underset{3}{r^8} + \underset{3}{r^3} + \underset{1}{r^9} + \underset{3}{r^{12}} + \underset{1}{r^5} + \underset{2}{r^{10}} + \underset{3}{r^4} +$$

$$\underset{3}{r^2} + \underset{2}{r^9} + \underset{4}{r^4} + \underset{3}{r^{10}} + \underset{2}{r^{13}} + \underset{4}{r^6} + \underset{2}{r^{11}} + \underset{2}{r^5} +$$

$$\underset{3}{r^{11}} + \underset{3}{r^1} + \underset{3}{r^{13}} + \underset{4}{r^2} + \underset{3}{r^5} + \underset{3}{r^{15}} + \underset{4}{r^3} + \underset{2}{r^{14}} +$$

$$\underset{3}{r^7} + \underset{3}{r^{14}} + \underset{3}{r^9} + \underset{4}{r^{15}} + \underset{4}{r^1} + \underset{4}{r^{11}} + \underset{3}{r^{16}} + \underset{4}{r^{10}} +$$

$$\underset{4}{r^5} + \underset{4}{r^{12}} + \underset{4}{r^7} + \underset{4}{r^{13}} + \underset{4}{r^{16}} + \underset{4}{r^9} + \underset{4}{r^{14}} + \underset{4}{r^8}$$

$$= -4.$$

**Fig. 16.3** Computation of $a_0 a_1$; below each root is the number of occurrences of the root so far

Since $a_0 + a_1 = -1$ and $a_0 a_1 = -4$, $a_1, a_2$ are the roots of the quadratic equation $y^2 + y - 4 = 0$ and they can be computed using the simple formula for the roots of a quadratic equation:

$$a_{0,1} = \frac{-1 \pm \sqrt{17}}{2} \, .$$

Now, let $b_0, b_1, b_2, b_3$ be the sums of every fourth root starting from $r^1, r^3, r^9, r^{10}$, respectively:

$$b_0 = r^1 + r^{13} + r^{16} + r^4$$
$$b_1 = r^3 + r^5 + r^{14} + r^{12}$$
$$b_2 = r^9 + r^{15} + r^8 + r^2$$
$$b_3 = r^{10} + r^{11} + r^7 + r^6 \, .$$

Check that $b_0 + b_2 = a_0$, $b_1 + b_3 = a_1$ and compute the corresponding products:

$$b_0 b_2 = (r + r^{13} + r^{16} + r^4) \times (r^9 + r^{15} + r^8 + r^2)$$
$$= (r^{10} + r^{16} + r^9 + r^3) + (r^5 + r^{11} + r^4 + r^{15}) +$$
$$(r^8 + r^{14} + r^7 + r^1) + (r^{13} + r^2 + r^{12} + r^6)$$
$$= -1 \, .$$

$$b_1 b_3 = (r^3 + r^5 + r^{14} + r^{12}) \times (r^{10} + r^{11} + r^7 + r^6)$$
$$= (r^{13} + r^{14} + r^{10} + r^9) + (r^{15} + r^{16} + r^{12} + r^{11}) +$$
$$(r^7 + r^8 + r^4 + r^3) \quad + (r^5 + r^6 + r^2 + r^1)$$
$$= -1.$$

To summarize these computations:

$$b_0 + b_2 = a_0$$
$$b_0 b_2 = -1$$
$$b_1 + b_3 = a_1$$
$$b_1 b_3 = -1,$$

so $b_0, b_2$ are the solutions of $y^2 - a_0 y - 1 = 0$, and $b_1, b_3$ are the solutions of $y^2 - a_1 y - 1 = 0$. Using the values previously computed for $a_0, a_1$ we can compute the roots $b_0, b_1$ (Fig. 16.4).

Finally, let $c_0, c_4$ be the sums of every eighth root starting with $r^1, r^{13}$:

$$c_0 = r^1 + r^{16}$$
$$c_4 = r^{13} + r^4$$
$$c_0 + c_4 = r^1 + r^{16} + r^{13} + r^4 = b_0$$
$$c_0 c_4 = (r^1 + r^{16}) \cdot (r^{13} + r^4)$$
$$= r^{14} + r^5 + r^{12} + r^3 = b_1,$$

so $c_0, c_4$ are the roots of $y^2 - b_0 y + b_1 = 0$. Since $\cos(360°/17) = c_0/2$ (Fig. 16.5) it suffices to compute the root $c_0 = r^1 + r^{16}$ (Fig. 16.6).

The cosine of the central angle of a heptadecagon is constructible with a straightedge and compass since it is composed only of rational numbers and the operations $\{+, -, \times, /, \sqrt{}\}$:

$$\cos\left(\frac{360°}{17}\right) = \frac{c_0}{2} \tag{16.4}$$

$$= -\frac{1}{16} + \frac{1}{16}\sqrt{17} + \frac{1}{16}\sqrt{34 - 2\sqrt{17}} + \tag{16.5}$$

$$\frac{1}{16}\sqrt{68 + 12\sqrt{17} + 2(-1 + \sqrt{17})\sqrt{34 - 2\sqrt{17}} - 16\sqrt{34 + 2\sqrt{17}}}. \tag{16.6}$$

$$b_0 = \frac{a_0 + \sqrt{a_0^2 + 4}}{2}$$

$$= \frac{\frac{(-1+\sqrt{17})}{2} + \sqrt{\left(\frac{(-1+\sqrt{17})}{2}\right)^2 + 4}}{2}$$

$$= \frac{(-1+\sqrt{17}) + \sqrt{\left(-1+\sqrt{17}\right)^2 + 16}}{4}$$

$$= \frac{(-1+\sqrt{17}) + \sqrt{34 - 2\sqrt{17}}}{4}$$

$$b_1 = \frac{a_1 + \sqrt{a_1^2 + 4}}{2}$$

$$= \frac{\frac{(-1-\sqrt{17})}{2} + \sqrt{\left(\frac{(-1-\sqrt{17})}{2}\right)^2 + 4}}{2}$$

$$= \frac{(-1-\sqrt{17}) + \sqrt{\left(-1-\sqrt{17}\right)^2 + 16}}{4}$$

$$= \frac{(-1-\sqrt{17}) + \sqrt{34 + 2\sqrt{17}}}{4} \, .$$

**Fig. 16.4** Computation of $b_0$ and $b_1$

$$r_1 + r_{16} = \cos\left(\frac{2\pi}{17}\right) + i\sin\left(\frac{2\pi}{17}\right) + \cos\left(\frac{2 \cdot 16\pi}{17}\right) + i\sin\left(\frac{2 \cdot 16\pi}{17}\right)$$

$$= \cos\left(\frac{2\pi}{17}\right) + i\sin\left(\frac{2\pi}{17}\right) + \cos\left(\frac{-2\pi}{17}\right) + i\sin\left(\frac{-2\pi}{17}\right)$$

$$= 2\cos\left(\frac{2\pi}{17}\right) \, .$$

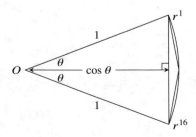

**Fig. 16.5** The cosine of the central angle computed from $r_1$, $r_{16}$

$$c_0 = \frac{b_0 + \sqrt{b_0^2 - 4b_1}}{2}$$

$$= \frac{1}{2}\frac{(-1 + \sqrt{17}) + \sqrt{34 - 2\sqrt{17}}}{4} +$$

$$\frac{1}{2}\sqrt{\left(\frac{(-1 + \sqrt{17}) + \sqrt{34 - 2\sqrt{17}}}{4}\right)^2 - 4\left(\frac{(-1 - \sqrt{17}) + \sqrt{34 + 2\sqrt{17}}}{4}\right)}$$

$$= -\frac{1}{8} + \frac{1}{8}\sqrt{17} + \frac{1}{8}\sqrt{34 - 2\sqrt{17}} +$$

$$\frac{1}{8}\sqrt{\left((-1 + \sqrt{17}) + \sqrt{34 - 2\sqrt{17}}\right)^2 - 16\left((-1 - \sqrt{17}) + \sqrt{34 + 2\sqrt{17}}\right)}$$

$$= -\frac{1}{8} + \frac{1}{8}\sqrt{17} + \frac{1}{8}\sqrt{34 - 2\sqrt{17}} +$$

$$\frac{1}{8}\sqrt{(-1 + \sqrt{17})^2 + 2(-1 + \sqrt{17})\sqrt{34 - 2\sqrt{17}} + (34 - 2\sqrt{17}) -}$$

$$\overline{\left((-16 - 16\sqrt{17}) + 16\sqrt{34 + 2\sqrt{17}}\right)}$$

$$= -\frac{1}{8} + \frac{1}{8}\sqrt{17} + \frac{1}{8}\sqrt{34 - 2\sqrt{17}} +$$

$$\frac{1}{8}\sqrt{68 + 12\sqrt{17} + 2(-1 + \sqrt{17})\sqrt{34 - 2\sqrt{17}} - 16\sqrt{34 + 2\sqrt{17}}}$$

**Fig. 16.6** Computation of $c_0$

## 16.5 Derivation of Gauss's Formula

The above formula for $\cos(360°/17)$ is not the one given by Gauss. Here is a derivation of Gauss's formula:

Let us simplify $2(-1 + \sqrt{17})\sqrt{34 - 2\sqrt{17}}$:

$$2(-1 + \sqrt{17})\sqrt{34 - 2\sqrt{17}} = -2\sqrt{34 - 2\sqrt{17}} + 2\sqrt{17}\sqrt{34 - 2\sqrt{17}}$$

$$+4\sqrt{34 - 2\sqrt{17}} - 4\sqrt{34 - 2\sqrt{17}}$$

$$= 2\sqrt{34 - 2\sqrt{17}} + 2\sqrt{17}\sqrt{34 - 2\sqrt{17}}$$

$$-4\sqrt{34 - 2\sqrt{17}}$$

$$= 2(1 + \sqrt{17})\sqrt{34 - 2\sqrt{17}} - 4\sqrt{34 - 2\sqrt{17}}.$$

We will remember the term $-4\sqrt{34-2\sqrt{17}}$ for now and simplify the first term by squaring it and then taking the square root:

$$2(1+\sqrt{17})\sqrt{34-2\sqrt{17}} = 2\sqrt{\left[(1+\sqrt{17})\sqrt{34-2\sqrt{17}}\right]^2}$$

$$= 2\sqrt{(18+2\sqrt{17})(34-2\sqrt{17})}$$

$$= 2\sqrt{(18\cdot34-4\cdot17)+\sqrt{17}(2\cdot34-2\cdot18)}$$

$$= 2\cdot4\sqrt{34+2\sqrt{17}}.$$

Substituting terms results in Gauss's formula:

$$\cos\left(\frac{360°}{17}\right) = -\frac{1}{16}+\frac{1}{16}\sqrt{17}+\frac{1}{16}\sqrt{34-2\sqrt{17}}$$

$$+\frac{1}{16}\sqrt{68+12\sqrt{17}+8\sqrt{34+2\sqrt{17}}-4\sqrt{34-2\sqrt{17}}-16\sqrt{34+2\sqrt{17}}}$$

$$= -\frac{1}{16}+\frac{1}{16}\sqrt{17}+\frac{1}{16}\sqrt{34-2\sqrt{17}}$$

$$+\frac{1}{8}\sqrt{17+3\sqrt{17}-\sqrt{34-2\sqrt{17}}-2\sqrt{34+2\sqrt{17}}}.$$

## 16.6 Construction of a Heptadecagon

Construct a unit circle centered at $O$ with perpendicular diameters $\overline{QP}$ and $\overline{SR}$ (Fig. 16.7). Construct $A$ so that $\overline{OA} = (1/4)\overline{OR}$.

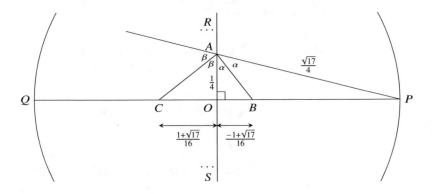

**Fig. 16.7** Construction of a heptadecagon (1)

By Pythagoras's Theorem:

$$\overline{AP} = \sqrt{\overline{OA}^2 + \overline{OP}^2} = \sqrt{(1/4)^2 + 1^2} = \sqrt{17}/4.$$

Let $B$ be the intersection of the internal bisector of $\angle OAP$ and the line segment $\overline{OP}$ and let $C$ be the intersection of the external bisector of $\angle OAP$ and the line segment $\overline{QO}$. By the internal angle bisector theorem (Thm. A.13):

$$\frac{\overline{OB}}{\overline{BP}} = \frac{\overline{AO}}{\overline{AP}}$$

$$\frac{\overline{OB}}{1 - \overline{OB}} = \frac{1/4}{\sqrt{17}/4}$$

$$\overline{OB} = \frac{1}{1 + \sqrt{17}} = \frac{1}{1 + \sqrt{17}} \cdot \frac{1 - \sqrt{17}}{1 - \sqrt{17}}$$

$$= \frac{-1 + \sqrt{17}}{16},$$

and by the external angle bisector theorem (Thm. A.14):

$$\frac{\overline{OC}}{\overline{CP}} = \frac{\overline{AO}}{\overline{AP}}$$

$$\frac{\overline{OC}}{1 + \overline{OC}} = \frac{1/4}{\sqrt{17}/4}$$

$$\overline{OC} = \frac{1}{-1 + \sqrt{17}} = \frac{1}{-1 + \sqrt{17}} \cdot \frac{1 + \sqrt{17}}{1 + \sqrt{17}}$$

$$= \frac{1 + \sqrt{17}}{16}.$$

Construct $D$ on $\overline{OP}$ such that $\overline{CD} = \overline{CA} = a$ (Fig. 16.8). By Pythagoras's Theorem:

$$\overline{CD} = \overline{CA} = \sqrt{\overline{OA}^2 + \overline{OC}^2}$$

$$= \sqrt{\left(\frac{1}{4}\right)^2 + \left(\frac{1 + \sqrt{17}}{16}\right)^2} = \frac{1}{16}\sqrt{16 + 1 + 17 + 2\sqrt{17}}$$

$$= \frac{1}{16}\sqrt{34 + 2\sqrt{17}}.$$

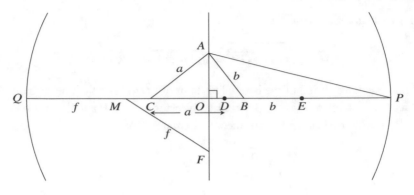

**Fig. 16.8** Construction of a heptadecagon (2)

Construct $E$ on $\overline{OP}$ such that $\overline{BE} = \overline{BA} = b$; again by Pythagoras's Theorem:

$$\overline{BE} = \overline{BA} = \sqrt{\overline{OA}^2 + \overline{OB}^2}$$

$$= \sqrt{\left(\frac{1}{4}\right)^2 + \left(\frac{-1+\sqrt{17}}{16}\right)^2} = \frac{1}{16}\sqrt{16 + 1 + 17 - 2\sqrt{17}}$$

$$= \frac{1}{16}\sqrt{34 - 2\sqrt{17}}.$$

Construct $M$ as the midpoint of $\overline{QD}$ and construct $F$ on $\overline{OS}$ such that $\overline{MF} = \overline{MQ} = f$:

$$\overline{MF} = \overline{MQ} = \frac{1}{2}\overline{QD} = \frac{1}{2}(\overline{QC} + \overline{CD}) = \frac{1}{2}((1 - \overline{OC}) + \overline{CD})$$

$$= \frac{1}{2}\left[1 - \left(\frac{1+\sqrt{17}}{16}\right) + \frac{\sqrt{34 + 2\sqrt{17}}}{16}\right]$$

$$= \frac{1}{32}\left(15 - \sqrt{17} + \sqrt{34 + 2\sqrt{17}}\right).$$

Note that $\overline{MO} = 1 - \overline{MQ} = 1 - \overline{MF}$.

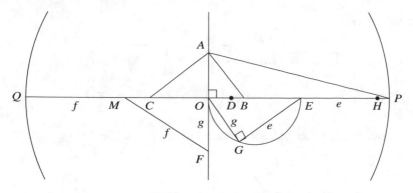

**Fig. 16.9** Construction of a heptadecagon (3)

Construct a semicircle whose diameter is $\overline{OE}$. Construct a chord $\overline{OG} = \overline{OF} = g$ (Fig. 16.9). By Pythagoras's Theorem:

$$
\begin{aligned}
\overline{OG} = \overline{OF} &= \sqrt{\overline{MF}^2 - \overline{MO}^2} = \sqrt{\overline{MF}^2 - (1 - \overline{MF})^2} \\
&= \sqrt{2\overline{MF} - 1} \\
&= \sqrt{\frac{1}{16}\left(15 - \sqrt{17} + \sqrt{34 + 2\sqrt{17}}\right) - 1} \\
&= \frac{1}{4}\sqrt{-1 - \sqrt{17} + \sqrt{34 + 2\sqrt{17}}}.
\end{aligned}
$$

$\angle OGE$ is a right angle since it is subtended by a diameter of the circle. Construct $H$ on $\overline{OP}$ such that $\overline{EH} = \overline{EG} = e$; again by Pythagoras's Theorem:

$$
\begin{aligned}
\overline{EH} = \overline{EG} &= \sqrt{\overline{OE}^2 - \overline{OG}^2} = \sqrt{(\overline{OB} + \overline{BE})^2 - \overline{OG}^2} \\
&= \sqrt{\left(\frac{-1 + \sqrt{17}}{16} + \frac{\sqrt{34 - 2\sqrt{17}}}{16}\right)^2 - \frac{1}{16}\left(-1 - \sqrt{17} + \sqrt{34 + 2\sqrt{17}}\right)} \\
&= \frac{1}{16}\sqrt{\left((18 - 2\sqrt{17}) + 2(-1 + \sqrt{17})\sqrt{34 - 2\sqrt{17}} + (34 - 2\sqrt{17})\right)} \\
&\qquad\qquad + \left(16 + 16\sqrt{17} - 16\sqrt{34 + 2\sqrt{17}}\right) \\
&= \frac{1}{16}\sqrt{68 + 12\sqrt{17} - 16\sqrt{34 + 2\sqrt{17}} - 2(1 - \sqrt{17})\sqrt{34 - 2\sqrt{17}}}.
\end{aligned}
$$

Compute $\overline{OE}$:

$$\overline{OE} = \overline{OB} + \overline{BE} = \frac{-1 + \sqrt{17}}{16} + \frac{1}{16}\sqrt{34 - 2\sqrt{17}}$$

$$= \frac{1}{16}\left(-1 + \sqrt{17} + \sqrt{34 - 2\sqrt{17}}\right).$$

Finally, $\overline{OH} = \overline{OE} + \overline{EH}$ which is Gauss's formula for $\cos(360°/17)$.

## 16.7 Construction of a Regular Pentagon

The complex fifth roots of unity are:

$$1 + i \cdot 0, \quad \frac{\sqrt{5} - 1}{4} \pm i\frac{\sqrt{10 + 2\sqrt{5}}}{4}, \quad \frac{-\sqrt{5} - 1}{4} \pm i\frac{\sqrt{10 - 2\sqrt{5}}}{4}.$$

### 16.7.1 Trigonometry

The central angle of a regular pentagon is $360°/5 = 72°$. Let us compute $\cos 36°$ using the trigonometric identities for $2\theta$ and $\theta/2$ (Thms. A.2.1, A.7):

$$0 = \cos 90° = \cos(72° + 18°) = \cos 2 \cdot 36° \cos 36°/2 - \sin 2 \cdot 36° \sin 36°/2$$

$$= (2\cos^2 36° - 1)\sqrt{\frac{1 + \cos 36°}{2}} - 2\sin 36° \cos 36°\sqrt{\frac{1 - \cos 36°}{2}}.$$

There is now only one angle in the formula; let $x = \cos 36°$. Then:

$$(2x^2 - 1)\sqrt{\frac{1 + x}{2}} = 2\sqrt{1 - x^2} \cdot x \cdot \sqrt{\frac{1 - x}{2}}$$

$$(2x^2 - 1)\sqrt{1 + x} = 2\sqrt{1 - x} \cdot \sqrt{1 + x} \cdot x \cdot \sqrt{1 - x}$$

$$2x^2 - 1 = 2x(1 - x)$$

$$4x^2 - 2x - 1 = 0.$$

Solving the quadratic equation gives a constructible value:

$$\cos 36° = \frac{1 + \sqrt{5}}{4}.$$

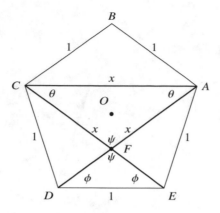

**Fig. 16.10** Construction of a regular pentagon (1)

## 16.7.2 Geometry

Let $\overline{ABCDE}$ be a regular pentagon (Fig. 16.10). By definition all the sides and all the interior angles are equal. It is easy to show by congruent triangles that all diagonals are equal. Let the length of the sides be 1 and the length of the diagonals be $x$.

$\triangle ACE \cong \triangle CAD$ by side-side-side so $\angle ACE = \angle CAD = \theta$. $\triangle AED \cong \triangle CDE$ by side-side-side so $\angle ADE = \angle CED = \phi$. $\angle AFC = \angle EFD = \psi$ are vertical angles. In both triangles the sum of the angles is 180° so $\psi + 2\theta = \psi + 2\phi$ and $\theta = \phi$. By alternate interior angles we conclude that $\overline{AC} \parallel \overline{DE}$.

Construct a line through $E$ parallel to $\overline{DC}$ and let $F$ be its intersection with $\overline{AC}$ (Fig. 16.11). $\overline{CDEF}$ is a rhombus so $\overline{EF} = \overline{CD} = \overline{AE} = 1$. $\triangle ACE$ is an isoceles triangle with base angles $\alpha$. $\triangle AEF$ is also isoceles and $\angle AFE = \angle FAE = \alpha$ so $\triangle ACE \sim \triangle AEF$. Taking ratios of the sides gives:

$$\frac{x}{1} = \frac{1}{x-1}.$$

The result is a quadratic equation $x^2 - x - 1 = 0$ whose positive root is constructible:

$$x = \frac{1 + \sqrt{5}}{2}.$$

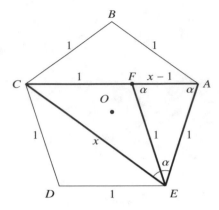

**Fig. 16.11** Construction of a regular pentagon (2)

## What Is the Surprise?

It is surprising that two millennia passed from the work of the Greeks on construction to the discovery by Gauss of the constructibility of the regular heptadecagon. It is also surprising that the problem was solved not by using geometry but by inventing new algebraic methods that had a far-reaching influence in mathematics.

## Sources

This chapter is based on [6]. Gauss's original work is available in an English translation [18]. Equation 16.5–16.6 appears in [41]; the author assigns an exercise to transform it into Gauss's formula as it appears in [18, p. 458] and [6, p. 68].

The construction of the heptadecagon is taken from [10] while other constructions can be found in [55]. The trigonometric construction of the regular pentagon is from [59]. The geometric construction of the regular pentagon was obtained by solving exercises 2.3.3 and 2.3.4 in [47].

# Appendix A
# Theorems From Geometry and Trigonometry

This appendix presents theorems in geometry and trigonometry that may not be familiar to the reader, as well as theorems that may be familiar but whose proofs are not. Section A.1 presents three formulas for computing the area of a triangle. Section A.2 proves trigonometric identities. Although the formulas and identities are mostly familiar, students frequently learn these identities by heart or look them up without ever seeing a proof. The following sections contain proofs of advanced theorems in geometry: Sect. A.3—the angle bisector theorems, Sect. A.4—Ptolemy's theorem that relates the sides and diagonals in a quadrilateral circumscribed by a circle, Sect. A.5—Ceva's theorem relating the three line segments of a triangle, and Sect. A.6—Menelaus's theorem on the segments of a transversal in a triangle.

## A.1 Theorems About Triangles

### A.1.1 Computing the Area of a Triangle

The standard formula for computing the area of a triangle from the base and the height is well-known. It can be proved using various geometric methods.

**Theorem A.1** *The area of the triangle $\triangle ABC$ is given by:*

$$\triangle ABC = \frac{1}{2}bh,\tag{A.1}$$

*where b, the base, is one of the sides of the triangle, and h, the height, is the length of the altitude to b from the opposite vertex (Fig. A.1a).*

**Proof** Figure A.1b shows that by "cutting" the triangle at half the height, we can "move" the shaded triangles to form a rectangle of the same area as the triangle. The rectangle's base is $b$ and its height is $h/2$. □

© The Author(s) 2022
M. Ben-Ari, *Mathematical Surprises*, https://doi.org/10.1007/978-3-031-13566-8

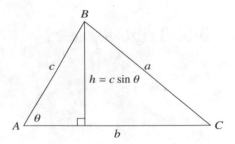

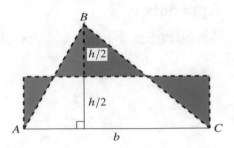

**Fig. A.1a** Computation of the area of a triangle from the base and the height

**Fig. A.1b** Computation of the area of a triangle from the base and the height

**Theorem A.2** *The area of the triangle $\triangle ABC$ is given by:*

$$\triangle ABC = \frac{1}{2}bc\sin\theta . \tag{A.2}$$

**Proof** From Thm. A.1 using $h = c\sin\theta$. □

**Theorem A.3 (Heron)** *The area of the triangle $\triangle ABC$ is given by:*

$$\triangle ABC = \sqrt{s(s-a)(s-b)(s-c)} ,$$

*where $s$, the* semi-perimeter *of the triangle, is equal to* $\frac{1}{2}(a+b+c)$.

**Proof** A radius of a circle and a tangent that intersects the radius are perdendicular. Furthermore, the lengths of the line segments of two tangents from the same point to the circle are equal. Therefore (Fig. A.2):[1]

$$\triangle AOB' \cong \triangle AOC', \quad \triangle BOA' \cong \triangle BOC', \quad \triangle COA' \cong \triangle COB' .$$

The area $\triangle ABC$ is the sum of the six triangles listed above. Since the height of six triangles is $r$, the radius of the inscribed circle, we obtain:

$$\triangle ABC = \triangle AOB' + \triangle AOC' + \triangle BOA' + \triangle BOC' + \triangle COA' + \triangle COB' \tag{A.3a}$$

$$\triangle ABC = \frac{1}{2}r(u+u+v+v+w+w) \tag{A.3b}$$

$$\triangle ABC = \frac{1}{2}r(a+b+c) \tag{A.3c}$$

$$\triangle ABC = rs . \tag{A.3d}$$

---

[1] This shows that the *incenter*, the center of the inscribed circle, is the common intersection of the three angle bisectors.

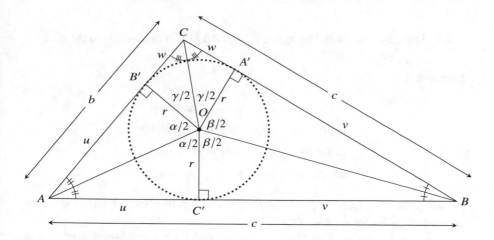

**Fig. A.2** Triangle with an inscribed circle

Let us now define the sides in terms of the tangents of the central angles:

$$\tan\frac{\alpha}{2} = \frac{u}{r}, \quad \tan\frac{\beta}{2} = \frac{v}{r}, \quad \tan\frac{\gamma}{2} = \frac{w}{r}.$$

From these definitions and $s = \frac{1}{2}(2u + 2u + 2w)$ we get:

$$s = u + v + w = r\left(\tan\frac{\alpha}{2} + \tan\frac{\beta}{2} + \tan\frac{\gamma}{2}\right).$$

Since $\frac{\alpha}{2} + \frac{\alpha}{2} + \frac{\beta}{2} + \frac{\beta}{2} + \frac{\gamma}{2} + \frac{\gamma}{2} = 360°$ and thus $\frac{\alpha}{2} + \frac{\beta}{2} + \frac{\gamma}{2} = 180°$, by Thm. A.11:

$$s = r\left(\tan\frac{\alpha}{2}\tan\frac{\beta}{2}\tan\frac{\gamma}{2}\right)$$

$$= r\left(\frac{u}{r}\frac{v}{r}\frac{w}{r}\right) = \frac{1}{r^2}(u\,v\,w)$$

$$r = \sqrt{\frac{u\,v\,w}{s}}.$$

By Eq. A.3d:

$$\triangle ABC = rs = s\sqrt{\frac{u\,v\,w}{s}} = \sqrt{s\,u\,v\,w}.$$

Heron's formula follows from $u = s - a, v = s - b, w = s - c$. $\qquad\qquad\qquad\square$

## A.2  Trigonometric Identities

### A.2.1  The Sine and Cosine of the Sum and Difference of Two Angles

**Theorem A.4**

$$\sin(\alpha + \beta) = \sin \alpha \cos \beta + \cos \alpha \sin \beta$$
$$\sin(\alpha - \beta) = \sin \alpha \cos \beta - \cos \alpha \sin \beta$$
$$\cos(\alpha + \beta) = \cos \alpha \cos \beta - \sin \alpha \sin \beta$$
$$\cos(\alpha - \beta) = \cos \alpha \cos \beta + \sin \alpha \sin \beta .$$

We will prove the first formula; the other formulas can be obtained using the values of sine and cosine for $-\alpha$ and $90° - \alpha$.

Given a right triangle $\triangle ABC$ with acute angle $\alpha$ and a right triangle $\triangle ACD$ with acute angle $\beta$, we can join them to obtain geometric figures with an angle $\alpha + \beta$ (Fig. A.3). The left diagram is the one most often used in proofs of the identities. Here we give two proofs based on the center and right diagrams.

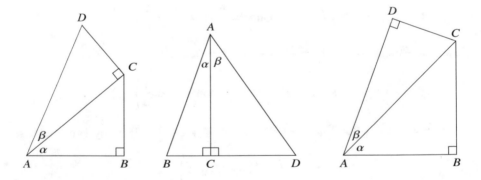

**Fig. A.3** Diagrams for proving the identity for the sine of sums of angles

***Proof (1)***  Let us compute the area of $\triangle ABD$ in two different ways: (1) using Eq. A.2 on $\triangle ABD$, and (2) using the equation separately on $\triangle ABC$ and $\triangle ADC$ (Fig. A.4). $h$ is also computed twice using the definition of the trigonometric functions:

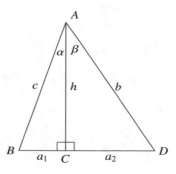

**Fig. A.4** Computation of the area of a triangle in two ways

$$\triangle ABD = \frac{1}{2}bc\sin(\alpha + \beta)$$

$$\triangle ABD = \triangle ABC + \triangle ADC$$

$$= \frac{1}{2}ch\sin\alpha + \frac{1}{2}bh\sin\beta$$

$$= \frac{1}{2}c(b\cos\beta)\sin\alpha + \frac{1}{2}b(c\cos\alpha)\sin\beta .$$

Equating the two formulas for $\triangle ABD$ and canceling $\frac{1}{2}bc$, we get:

$$\sin(\alpha + \beta) = \sin\alpha\cos\beta + \cos\alpha\sin\beta .$$ $\square$

The second proof uses the following theorem:

**Theorem A.5** *In a circle of* diameter 1 *the length of a chord that subtends an inscribed angle is equal to the sine of the angle (Fig. A.5).*

***Proof*** Let $\overline{AB}$ be a diameter and let $\angle BAC = \alpha$. Let $D$ be any other point on the circle one of whose sides is the chord $\overline{BC}$. Since equal chords subtend equal inscribed angles $\angle BDC = \alpha$. In the right triangle $\triangle ABC$:

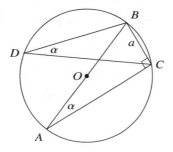

**Fig. A.5** All inscribed angles subtended by a chord are equal

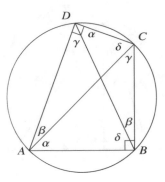

**Fig. A.6** A quadrilateral circumscribed by a circle

$$\sin \alpha = \frac{\overline{BC}}{\overline{AB}} = \frac{\overline{BC}}{1} = \overline{BC}.$$

$\square$

**Proof (2)** This proof is based on the right diagram in Fig. A.3 reproduced in Fig. A.6, where the quadrilateral $\overline{ABCD}$ has been inscribed in a circle. By Thm. A.15 a quadrilateral can be circumscribed by a circle if and only if the sum of each pair of opposite angles is 180°. $\angle ADC + \angle ABC = 180°$ since both angles are right angles. From Thm. 5.4 the sum of the interior angles of a quadrilateral is 360°, so $\angle DAB + \angle DCB = 180°$.

Let the diameter of the circle be 1 (otherwise, multiply everything by the length of the diameter). Then the sides of the quadrilateral are:

$$\overline{BC} = \sin \alpha, \quad \overline{CD} = \sin \beta, \quad \overline{AB} = \sin \gamma, \quad \overline{DA} = \sin \delta,$$

and their diagonals are:

$$\overline{BD} = \sin(\alpha + \beta), \quad \overline{CA} = \sin(\alpha + \gamma).$$

By Ptolemy's Theorem (Thm. A.18) the product of the diagonals of a quadrilateral circumscribed by a circle is equal to the sum of the products of opposite sides of the quadrilateral. Since $\angle ADC$ and $\angle ABC$ are right angles we have:

$$\sin(\alpha + \beta) \sin(\alpha + \gamma) = \sin \alpha \sin \delta + \sin \beta \sin \gamma$$

$$\sin(\alpha + \beta) \sin(90°) \quad = \sin \alpha \sin(90° - \beta) + \sin \beta \sin(90° - \alpha)$$

$$\sin(\alpha + \beta) \quad\quad\quad\quad = \sin \alpha \cos \beta + \cos \alpha \sin \beta.$$

$\square$

## A.2.2 The Cosine of a Triple Angle

**Theorem A.6**

$$\cos 3\alpha = 4\cos^3 \alpha - 3\cos \alpha.$$

***Proof*** The proof uses the formulas in Thm. A.4 and the formula $\sin^2 \alpha + \cos^2 \alpha = 1$:

$$
\begin{aligned}
\cos 3\alpha &= \cos(2\alpha + \alpha) \\
&= \cos 2\alpha \cos \alpha - \sin 2\alpha \sin \alpha \\
&= (\cos^2 \alpha - \sin^2 \alpha)\cos \alpha - (2\sin \alpha \cos \alpha)\sin \alpha \\
&= \cos^3 \alpha - \cos \alpha \sin^2 \alpha - 2\sin^2 \alpha \cos \alpha) \\
&= \cos^3 \alpha - \cos \alpha + \cos^3 \alpha - 2\cos \alpha + 2\cos^3 \alpha \\
&= 4\cos^3 \alpha - 3\cos \alpha.
\end{aligned}
$$

□

## A.2.3 The Sine and Cosine of a Half-Angle

**Theorem A.7** *If $\alpha$ is an angle in a* triangle *then:* [2]

$$
\cos\left(\frac{\alpha}{2}\right) = \sqrt{\frac{1 + \cos \alpha}{2}}
$$

$$
\sin\left(\frac{\alpha}{2}\right) = \sqrt{\frac{1 - \cos \alpha}{2}}.
$$

***Proof*** The proof uses the formulas Thm. A.4 and the formula $\sin^2 \alpha + \cos^2 \alpha = 1$:

$$
\cos \alpha = \cos 2\left(\frac{\alpha}{2}\right) = \cos\left(\frac{\alpha}{2}\right)\cos\left(\frac{\alpha}{2}\right) - \sin\left(\frac{\alpha}{2}\right)\sin\left(\frac{\alpha}{2}\right)
$$

$$
= 2\cos^2\left(\frac{\alpha}{2}\right) - 1
$$

$$
\cos\left(\frac{\alpha}{2}\right) = \sqrt{\frac{1 + \cos \alpha}{2}}
$$

$$
\sin^2\left(\frac{\alpha}{2}\right) = 1 - \cos^2\left(\frac{\alpha}{2}\right) = 1 - \frac{1 + \cos \alpha}{2}
$$

$$
\sin\left(\frac{\alpha}{2}\right) = \sqrt{\frac{1 - \cos \alpha}{2}}.
$$

□

---

[2] The general formula is more complex because the square roots can be either positive or negative depending on the quadrant in which $\alpha/2$ is located. For a triangle $0 < \alpha < 180°$, so $0 < \alpha/2 < 90°$ is in the first quadrant and both the sine and the cosine are positive.

### A.2.4 The Law of Cosines

**Theorem A.8 (Law of cosines)** *In a triangle $\triangle ABC$ with sides $a, b, c$ (Fig. A.7):*

$$c^2 = a^2 + b^2 - 2ab \cos \angle ACB.$$

***Proof (1)*** Drop an an altitude from $C$ to $\overline{AB}$ and use the definition of cosine and Pythagoras's Theorem:

$$c = x + (c - x) = a \cos \beta + b \cos \alpha \tag{A.4a}$$

$$c^2 = ac \cos \beta + bc \cos \alpha. \tag{A.4b}$$

Similarly, drop altitudes from $A$ to $\overline{BC}$ and from $B$ to $\overline{AC}$ to obtain:

$$a^2 = ca \cos \beta + ba \cos \gamma \tag{A.5a}$$

$$b^2 = cb \cos \alpha + ab \cos \gamma. \tag{A.5b}$$

Adding Eqs. A.5a and A.5b and subtracting Eq. A.4b gives:

$$
\begin{aligned}
a^2 + b^2 - c^2 &= ca \cos \beta + ba \cos \gamma \\
&\quad + cb \cos \alpha + ab \cos \gamma \\
&\quad - ac \cos \beta - bc \cos \alpha \\
&= 2ab \cos \gamma \\
c^2 &= a^2 + b^2 - 2ab \cos \gamma.
\end{aligned}
$$

$\square$

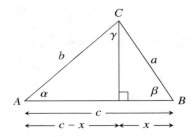

**Fig. A.7** Proof 1 of the Law of Cosines

***Proof (2)*** The second proof uses Ptolemy's theorem (Thm. A.18).[3]

The triangle $\triangle ABC$ can be circumscribed by a circle. Construct another triangle $\triangle ABC'$ congruent with $\triangle ABC$ and inscribed within the same circle (Fig. A.8). This can be done by constructing an angle from $\overline{AB}$ equal to $\angle CAB$ which intersects the circle at $C'$ and then constructing the line $\overline{C'A}$. Since angles that are subtended by the same chord are equal $\angle AC'B = \angle BCA$, so also $\angle CBA = \angle C'AB$ and thus $\triangle ABC' \cong \triangle BAC$ by angle-side-angle with the common side $\overline{AB}$.

Drop perpendiculars from $C$ to $D$ and from $C'$ to $D'$ on $\overline{AB}$ so that $x = a\cos\beta$. By Ptolemy's theorem for the quadrilateral $\overline{ABCC'}$:

$$\begin{aligned} b^2 &= a^2 + c(c - 2x) \\ &= a^2 + c(c - 2a\cos\beta) \\ &= a^2 + c^2 - 2ac\cos\beta \,. \end{aligned} \qquad \square$$

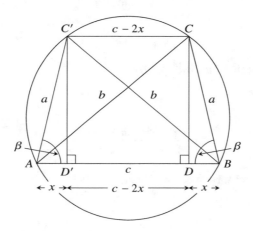

**Fig. A.8** Proof 2 of the Law of Cosines

---

[3] Section A.4 uses the Law of Cosines to prove Ptolemy's theorem! The first proof of the Law of Cosines avoids this circular reasoning. Furthermore, there are proofs of Ptolemy's theorem that do not use the Law of Cosines.

## A.2.5 The Tangent of the Sum of Two Angles

**Theorem A.9**

$$\tan(\alpha + \beta) = \frac{\tan\alpha + \tan\beta}{1 - \tan\alpha\tan\beta}.$$

*Proof*

$$
\begin{aligned}
\tan(\alpha + \beta) &= \frac{\sin(\alpha + \beta)}{\cos(\alpha + \beta)} \\
&= \frac{\sin\alpha\cos\beta + \cos\alpha\sin\beta}{\cos\alpha\cos\beta - \sin\alpha\sin\beta} \\
&= \frac{\sin\alpha + \cos\alpha\tan\beta}{\cos\alpha - \sin\alpha\tan\beta} \\
&= \frac{\tan\alpha + \tan\beta}{1 - \tan\alpha\tan\beta}.
\end{aligned}
$$

□

## A.2.6 The Tangent of a Half-Angle

**Theorem A.10**

$$\tan\left(\frac{\alpha}{2}\right) = \frac{-1 \pm \sqrt{1 + \tan^2\alpha}}{\tan\alpha}.$$

*Proof* We derive and solve a quadratic equation in $\tan\left(\frac{\alpha}{2}\right)$:

$$\tan\alpha = \frac{\tan\left(\frac{\alpha}{2}\right) + \tan\left(\frac{\alpha}{2}\right)}{1 - \tan\left(\frac{\alpha}{2}\right)\tan\left(\frac{\alpha}{2}\right)}$$

$$\tan\alpha\tan^2\left(\frac{\alpha}{2}\right) + 2\tan\left(\frac{\alpha}{2}\right) - \tan\alpha = 0$$

$$\tan\left(\frac{\alpha}{2}\right) = \frac{-1 \pm \sqrt{1 + \tan^2\alpha}}{\tan\alpha}.$$

□

### A.2.7 The Product of Three Tangents

**Theorem A.11** *If $\alpha + \beta + \gamma = 180°$ then:*

$$\tan \alpha + \tan \beta + \tan \gamma = \tan \alpha \tan \beta \tan \gamma .$$

*Proof*

$$\begin{aligned}
\tan \gamma &= \tan(180° - (\alpha + \beta)) \\
&= -\tan(\alpha + \beta) \\
&= -\frac{\tan \alpha + \tan \beta}{1 - \tan \alpha \tan \beta} \\
\tan \alpha \tan \beta \tan \gamma &= \tan \alpha + \tan \beta + \tan \gamma .
\end{aligned}$$

□

### A.2.8 The Limit of $\sin \alpha / \alpha$

**Theorem A.12**

$$\lim_{\alpha \to 0} \frac{\sin \alpha}{\alpha} = 1 .$$

*Proof* By examining regular polygons inscribed within a circle (Fig. A.9), we see that the more sides that a polygon has, the closer its perimeter is to the circumference of the circle. The circumference of the circle divided by the number of sides is the length of an arc with the same endpoints as the corresponding side, since in a regular polygon all sides have the same length. Since the ratio of the circumference of the circle to the perimeter of an inscribed polygon approaches 1 as the number of sides increases, so does the ratio of the length of an arc to the corresponding chord. This is demonstrated by the following numerical examples:

**Fig. A.9** Regular polygons with 3, 8 and 16 sides inscribed within a circle

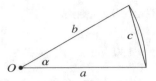

**Fig. A.10** The length of a chord corresponding to an arc of size $\alpha$

| Angle | Arc length | Chord length | Ratio |
|-------|-----------|--------------|-------|
| 80 | 1.396 | 1.286 | 1.090 |
| 60 | 1.047 | 1.000 | 1.047 |
| 40 | 0.698 | 0.684 | 1.006 |
| 5 | 0.087 | 0.087 | 1.000 |

Since $a = b = 1$ the length of the chord $c$ subtending $\alpha$ can be computed from the Law of Cosines (Fig. A.10):

$$c^2 = a^2 + b^2 - 2ab \cos \alpha$$
$$c = \sqrt{2 - 2 \cos \alpha}$$
$$\lim_{\alpha \to 0} c = \sqrt{2 - 2 \cdot 1} = 0.$$

Referring to Fig. A.11:

$$\lim_{\alpha \to 0} \frac{\sin \alpha}{\alpha} = \lim_{\alpha \to 0} \frac{2 \sin \alpha}{2\alpha}.$$

This is the ratio of the length of chord $\overline{PQ}$ to the length of arc $\overset{\frown}{PQ}$. But we have seen that this ratio converges to 1 as the subtended angle $2\alpha$ tends to 0, so:

$$\lim_{\alpha \to 0} \frac{\sin \alpha}{\alpha} = 1.$$

$\square$

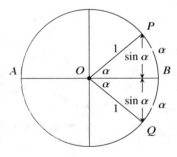

**Fig. A.11** Ratio of $\sin x$ to $x$

## A.3 The Angle Bisector Theorems

**Theorem A.13** *In* $\triangle ABC$ *let the angle bisector of* $\angle BAC$ *intersect* $\overline{BC}$ *at* $D$ *(Fig. A.12). Then:*

$$\frac{\overline{BD}}{\overline{CD}} = \frac{\overline{AB}}{\overline{AC}} .$$

*Proof* We prove the theorem by computing the areas of two triangles using both the base and height (Eq. A.1), and the base, angle and side (Eq. A.2):

$$\triangle ABD = \frac{1}{2}\overline{BD}h = \frac{1}{2}\overline{AB}\,\overline{AD}\sin\alpha$$

$$\frac{\overline{BD}}{\overline{AB}} = \frac{\overline{AD}\sin\alpha}{h}$$

$$\triangle ACD = \frac{1}{2}\overline{CD}h = \frac{1}{2}\overline{AC}\,\overline{AD}\sin\alpha$$

$$\frac{\overline{CD}}{\overline{AC}} = \frac{\overline{AD}\sin\alpha}{h}$$

$$\frac{\overline{BD}}{\overline{CD}} = \frac{\overline{AB}}{\overline{AC}} . \qquad\qquad \square$$

There is also an angle bisector theorem for the *external bisector*:

**Theorem A.14** *In* $\triangle ABC$ *let* $\overline{AE}$ *be the bisector of the angle supplementary to the angle* $\triangle BAC$ *(Fig. A.13) and let the bisector intersect* $\overline{BC}$ *at* $E$ *(Fig. A.12). Then:*

$$\frac{\overline{BE}}{\overline{CE}} = \frac{\overline{AB}}{\overline{AC}} .$$

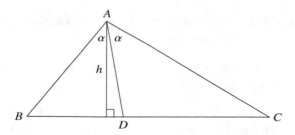

**Fig. A.12** The internal angle bisector theorem

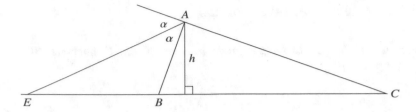

**Fig. A.13** The external angle bisector theorem

***Proof*** Since $\overline{AC}$ is a straight line $\angle EAC = 180° - \alpha$.

$$\triangle ABE = \frac{1}{2}\overline{BE}h = \cdot\frac{1}{2}\overline{AE}\,\overline{AB}\sin\alpha$$

$$\triangle ACE = \frac{1}{2}\overline{CE}h = \frac{1}{2}\overline{AE}\,\overline{AC}\sin(180° - \alpha) = \frac{1}{2}\overline{AE}\,\overline{AC}\sin\alpha$$

$$\frac{\overline{BE}}{\overline{AB}} = \frac{\overline{AE}\sin\alpha}{h} = \frac{\overline{CE}}{\overline{AC}}$$

$$\frac{\overline{BE}}{\overline{CE}} = \frac{\overline{AB}}{\overline{AC}}.$$

$\square$

## A.4 Ptolemy's Theorem

### A.4.1 A Trapezoid Circumscribed by a Circle

Before giving the proof of Ptolemy's theorem we prove theorems on quadrilaterals and trapezoids.

**Theorem A.15** *A quadrilateral can be circumscribed by a circle if and only if the opposite angles are supplementary (sum to* $180°$*).*

Geometry textbooks give the simple proof of the forward direction, but it is hard to find a proof of the converse so both proofs are given here.

***Proof (Forward direction)*** An inscribed angle is equal to half the arc that subtends it so $\angle DAB$ is half of the arc $\overset{\frown}{DCB}$ and $\angle DCB$ is half of the arc $\overset{\frown}{DAB}$ (Fig. A.14a). The two arcs form the entire circumference of the circle so their sum is $360°$. Therefore, $\angle DAB + \angle DCB = \frac{1}{2} \cdot 360° = 180°$, and similarly $\angle ADC + \angle ABC = 180°$.

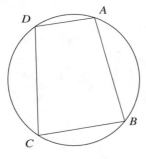

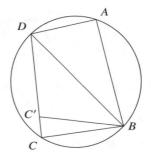

**Fig. A.14a** A quadrilateral circumscribed by a circle

**Fig. A.14b** The fourth vertex must be on the circumference

***Proof (Converse direction)*** Any triangle can be circumscribed by a circle. Circumscribe $\triangle DAB$ by a circle and suppose that $C'$ is a point such that $\angle DAB + \angle DC'B = 180°$, but $C'$ is *not* on the circumference of the circle. Without loss of generality, let $C'$ be within the circle (Fig. A.14b).

Construct a ray that extends $\overline{DC'}$ and let $C$ be its intersection with the circle. $\overline{ABCD}$ is circumscribed by a circle so:

$$\angle DAB + \angle DCB = 180° = \angle DAB + \angle DC'B$$
$$\angle DCB = \angle DC'B,$$

which is impossible if $C$ is on the circle and $C'$ is inside the circle.                    □

**Theorem A.16** *The opposite angles of an isosceles trapezoid are supplementary.*

***Proof*** Construct the line $\overline{AB'}$ parallel to $\overline{CD}$ (Fig. A.15). $\overline{AB'CD}$ is a parallelogram and $\triangle ABB'$ is an isosceles triangle, so $\angle C = \angle ABB' = \angle AB'B = \angle B$. Similarly, $\angle A = \angle D$. Since the sum of the internal angles of any quadrilateral is equal to $360°$:

$$\angle A + \angle B + \angle C + \angle D = 360°$$
$$2\angle A + 2\angle C = 360°$$
$$\angle A + \angle C = 180°,$$

and similarly $\angle B + \angle D = 180°$.                    □

**Theorem A.17** *An isoceles trapezoid can be be circumscribed by a circle.*

The proof is immediate by Thms. A.15, A.16.

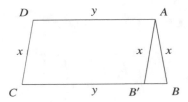

**Fig. A.15**  An isoceles trapezoid

## A.4.2  Proof of Ptolemy's Theorem

**Theorem A.18 (Ptolemy)** *Given a quadrilateral circumscribed by a circle, the following formula relates the lengths of the diagonals and the lengths of the sides (Fig. A.16).*

$$ef = ac + bd \, .$$

***Proof***  By the Law of Cosines for the four triangles $\triangle ABC$, $\triangle ADC$, $\triangle DAB$, $\triangle DCB$:

$$e^2 = a^2 + b^2 - 2ab \cos \angle B$$
$$e^2 = c^2 + d^2 - 2cd \cos \angle D$$
$$f^2 = a^2 + d^2 - 2ad \cos \angle A$$
$$f^2 = b^2 + c^2 - 2bc \cos \angle C \, .$$

$\angle C = 180° - \angle A$ and $\angle D = 180° - \angle B$ because they are opposite angles of a quadrilateral circumscribed by a circle, so $\cos \angle D = - \cos \angle B$ and $\cos \angle C = - \cos \angle A$. Eliminate the cosine term from the above equations to obtain:

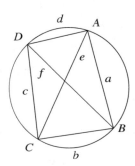

**Fig. A.16**  Ptolemy's theorem

$$e^2(cd + ab) = abc^2 + abd^2 + a^2cd + b^2cd$$
$$e^2 = \frac{(ac + bd)(ad + bc)}{(ab + cd)}$$
$$f^2 = \frac{(ab + cd)(ac + bd)}{(ad + bc)}.$$

Multiply the two equations and simplify to get Ptolemy's theorem:

$$e^2 \cdot f^2 = (ac + bd)^2$$
$$ef = (ac + bd).$$                                                                     □

## A.5 Ceva's Theorem

**Theorem A.19 (Ceva)** *Given line segments from the vertices of a triangle to the opposite edges that intersect in a point, the lengths of the segments satisfy (Fig. A.17):*

$$\frac{\overline{AM}}{\overline{MB}} \cdot \frac{\overline{BQ}}{\overline{QS}} \cdot \frac{\overline{SP}}{\overline{PA}} = 1.$$

**Proof** If the altitudes of two triangles are equal, their areas are proportional to the bases. In both diagrams in Fig. A.18, the altitudes of the gray triangles are equal, so:

$$\frac{\triangle BQO}{\triangle SQO} = \frac{\overline{BQ}}{\overline{QS}}, \qquad \frac{\triangle BQA}{\triangle SQA} = \frac{\overline{BQ}}{\overline{QS}}.$$

By subtracting the areas of the indicated triangles, we get the proportion between the gray triangles shown in Fig. A.19:

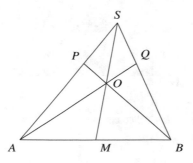

**Fig. A.17** Ceva's theorem

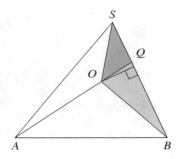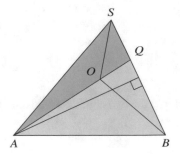

**Fig. A.18** Triangles in Ceva's theorem

$$\frac{\triangle BOA}{\triangle SOA} = \frac{\triangle BQA - \triangle BQO}{\triangle SQA - \triangle SQO} = \frac{\overline{BQ}}{\overline{QS}} \,.$$

This might look strange at first so we explain it using a simpler notation:

$$\frac{c}{d} = \frac{a}{b}$$
$$\frac{e}{f} = \frac{a}{b}$$
$$c - e = \frac{ad}{b} - \frac{af}{b} = \frac{a}{b}(d - f)$$
$$\frac{c - e}{d - f} = \frac{a}{b} \,.$$

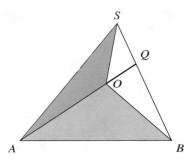

**Fig. A.19** Subtracting areas in Ceva's theorem

Similarly, we can prove:

$$\frac{\overline{AM}}{\overline{MB}} = \frac{\triangle AOS}{\triangle BOS}$$

$$\frac{\overline{SP}}{\overline{PA}} = \frac{\triangle SOB}{\triangle AOB},$$

so:

$$\frac{\overline{AM}}{\overline{MB}}\frac{\overline{BQ}}{\overline{QS}}\frac{\overline{SP}}{\overline{PA}} = \frac{\triangle AOS}{\triangle BOS}\frac{\triangle BOA}{\triangle SOA}\frac{\triangle SOB}{\triangle AOB} = 1,$$

since the order of the vertices in a triangle makes no difference.            □

## A.6 Menelaus's Theorem

### Theorem A.20 (Menelaus)

Let $\triangle ABC$ be a triangle and $\overline{DBQ}$ a transversal line *that intersects all three of the edges of the triangle or their extensions (Fig. A.20). Then:* [4]

$$\frac{\overline{AB}}{\overline{BP}} \cdot \frac{\overline{PQ}}{\overline{QC}} \cdot \frac{\overline{CD}}{\overline{AD}} = 1. \tag{A.6}$$

***Proof*** Draw a line through $C$ parallel to $\overline{AB}$ and extend $\overline{DQ}$ until it intersects the parallel at $K$. From $\triangle ADB \sim \triangle CDK$ it follows that:

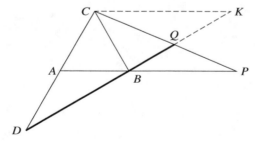

**Fig. A.20** Menelaus's theorem

---

[4] Depending on the configuration of the triangle and the transversal line, the result of the multiplication can be either plus or minus one.

$$\frac{\overline{CD}}{\overline{AD}} = \frac{\overline{CK}}{\overline{AB}} .$$

From $\triangle BQP \sim \triangle KQC$ it follows that:

$$\frac{\overline{QC}}{\overline{PQ}} = \frac{\overline{CK}}{\overline{BP}} .$$

Eliminating $\overline{CK}$ gives $\overline{AB} \cdot \overline{CD} \cdot \overline{PQ} = \overline{QC} \cdot \overline{BP} \cdot \overline{AD}$ which can be re-arranged to obtain Thm. A.6.                                                                                                □

## Sources

The appendix is based primarily on [19]. Ceva's theorem and Menelaus's theorem can be proved from each other [45].

# References

Wikipedia references are listed by their title only.

All links were alive on 7 December 2021.

1. Aigner, M., Ziegler, G.M.: Proofs from THE BOOK (Fifth Edition). Springer (2014)
2. Alperin, R.C.: A mathematical theory of origami constructions and numbers. New York Journal of Mathematics **6**, 119–133 (2000)
3. Barabash, M.: A non-visual counterexample in elementary geometry. The College Mathematics Journal **36**(5) (2005)
4. Ben-Ari, M.: Mathematical Logic for Computer Science (Third Edition). Springer (2012)
5. Ben-Ari, M.: LearnSAT: A SAT solver for education. Journal of Open Source Software **3**(24), 639 (2018). `https://doi.org/10.21105/joss.00639`
6. Bewersdorff, J.: Galois Theory for Beginners: A Historical Perspective. American Mathematical Society (2006)
7. Bold, B.: Famous Problems of Mathematics: A History of Constructions with Straight Edge and Compass. Van Nostrand (1969)
8. Bradford, P.V.: Visualizing solutions to n-th degree algebraic equations using right-angle geometric paths. Archived May 2, 2010 at the Wayback Machine. `https://web.archive.org/web/20100502013959/http://www.concentric.net/~pvb/ALG/rightpaths.html` (2010)
9. Butler IV, L.: Ramsey theory. `https://www.whitman.edu/Documents/Academics/Mathematics/2016/Barton.pdf` (2016)
10. Callagy, J.J.: The central angle of the regular 17-gon. The Mathematical Gazette **67**(442), 290–292 (1983). `https://www.jstor.org/stable/3617271`
11. Courant, R., Robbins, H.: What is Mathematics? An Elementary Approach to Ideas and Methods (Second Edition). Oxford University Press (1996). Revised by Ian Stewart
12. Davies, R.: On Langford's problem (II). The Mathematical Gazette **43**, 253–5 (1959)
13. Dörrie, H.: 100 Problems of Elementary Mathematics: Their History and Solution. Dover (1965)
14. Dörrie, H.: 100 problems of elementary mathematics: Their history and solution. Reworked by Michael Woltermann. Archived 21 February 2020 at the

© The Author(s) 2022

M. Ben-Ari, *Mathematical Surprises*, https://doi.org/10.1007/978-3-031-13566-8

Wayback Machine. https://web.archive.org/web/20191223032114/http://www2.washjeff.edu/users/mwoltermann/Dorrie/DorrieContents.htm (2010)

15. Dudley, U.: A Budget of Trisections. Springer (1987)
16. Eppstein, D.: Twenty proofs of Euler's formula: $V - E + F = 2$. https://www.ics.uci.edu/~eppstein/junkyard/euler/ (n.d.)
17. Fraleigh, J.B.: A First Course in Abstract Algebra (Seventh Edition). Addison-Wesley (2003)
18. Gauss, K.F.: Disquisitiones Arithmeticae. Yale University Press (2006). Editors: Todd W. Bressi and Paul Groth
19. Gelfand, I., Saul, M.: Trigonometry. Springer (2001)
20. Graham, R., Butler, S.: Rudiments of Ramsey Theory (Second Edition). American Mathematical Society (2015)
21. Gunderson, D.S.: Handbook of Mathematical Induction: Theory and Applications. Mathematical Association of America (2010)
22. Heath. T.L.: The Thirteen Books of Euclid's Elements. Dover (1956)
23. Heule, M.J.H., Kullmann, O.: The science of brute force. Communications of the ACM **60**(8), 70–79 (2017)
24. Hull, T.C.: Solving cubics with creases: The work of Beloch and Lill. American Mathematical Monthly **118**(4), 307–315 (2011)
25. Hungerbühler, N.: A short elementary proof of the Mohr-Mascheroni theorem. American Mathematical Monthly **101**(8), 784–787 (1994)
26. Lang, R.J.: Origami and geometric constructions. http://langorigami.com/wp-content/uploads/2015/09/origami_constructions.pdf (1996—2015)
27. Laugwitz, D.: Eine elementare Methode für Unmöglichkeitsbeweise bei Konstruktionen mit Zirkel und Lineal. Elemente der Mathematik **17**, 54–58 (1962)
28. Loh, P.S.: A different way to solve quadratic equations. https://www.poshenloh.com/quadratic/ (2019)
29. Loh, P.S.: A simple proof of the quadratic formula. https://arxiv.org/abs/1910.06709 (2019)
30. Manna, Z.: Mathematical Theory of Computing. McGraw-Hill (1974)
31. Martin. G.E.: Geometric Constructions. Springer (1998)
32. Mastin. L.: Muhammad ibn Musa al-Khwarizmi: Muslim Mathematician. https://www.storyofmathematics.com/islamic_alkhwarizmi.html (2020)
33. McCallum, W.: A tale of two triangles: Heron triangles and elliptic curves. http://blog.kleinproject.org/?p=4 (2012)
34. McKay, B.D.: Ramsey theory. http://users.cecs.anu.edu.au/~bdm/data/ramsey.html (nd)
35. Miller. J.: Langford's problem, remixed. http://dialectrix.com/langford.html (2014)
36. Newton. L.: The power of origami. https://plus.maths.org/content/power-origami
37. Peil. T.: The rusty compass theorem. Archived 20/07/2020 at the Wayback Machine. https://web.archive.org/web/20200720195718/http://web.mnstate.edu/peil/geometry/C2EuclidNonEuclid/1Compass.htm (2006)
38. Ramanujan: Squaring the circle. Journal of the Indian Mathematical Society **V**, 138 (1913). http://ramanujan.sirinudi.org/Volumes/published/ram05.pdf
39. Ramanujan: Modular equations and approximations to $\pi$. The Quarterly Journal of Pure and Applied Mathematics **XLV**, 350–372 (1914). http://ramanujan.sirinudi.org/Volumes/published/ram06.pdf
40. Riaz. M.: Geometric solutions of algebraic equations. American Mathematical Monthly **69**(7), 654–658 (1962)
41. Rike. T.: Fermat numbers and the heptadecagon. https://mathcircle.berkeley.edu/sites/default/files/BMC6/ps0506/Heptadecagon.pdf (2005)

42. Robson, E.: Words and pictures: New light on Plimpton 322. American Mathematical Monthly **109**(2), 105–120 (2002)
43. Ross, S.: A First Course in Probability (Tenth Edition). Pearson (2019)
44. Schumer, P.: The Josephus problem: Once more around. Mathematics Magazine **75**(1), 12–17 (2002)
45. Silvester, J.R.: Ceva = (Menelaus)$^2$. The Mathematical Gazette **84**(500), 268–271 (2000)
46. Sipka, T.: Alfred Bray Kempe's "Proof" of the four-color theorem. Math Horizons **10**(2), 21–26 (2002). http://www.jstor.org/stable/25678395
47. Stillwell, J.: Mathematics and Its History (Third Edition). Springer (2010)
48. Suzuki, J.: A brief history of impossibility. Mathematics Magazine **81**(1), 27–38 (2008)
49. Thomas, R.: An update on the four-color theorem. Notices of the AMS **45**(7), 848–859 (1998). http://www.ams.org/notices/199807/thomas.pdf
50. Toussaint, G.: A new look at Euclid's second proposition. The Mathematical Intelligencer **15**(3), 12–23 (1993)
51. Wikipedia: Angle trisection
52. Wikipedia: Cubic equation
53. Wikipedia: Five color theorem
54. Wikipedia: Four color theorem
55. Wikipedia: Heptadecagon
56. Wikipedia: Huzita–Hatori axioms
57. Wikipedia: Josephus problem
58. Wikipedia: Neusis construction
59. Wikipedia: Pentagon
60. Wikipedia: Plimpton 322
61. Wikipedia: Quadratic equation
62. Wikipedia: Quadratrix of Hippias
63. Wikipedia: Sexagesimal

# Index

© The Author(s) 2022
M. Ben-Ari, *Mathematical Surprises*, https://doi.org/10.1007/978-3-031-13566-8

Printed in the United States
by Baker & Taylor Publisher Services